GEL ELECTROPHORESIS: PROTEINS

The INTRODUCTION TO BIOTECHNIQUES series

Editors:

J.M. Graham MIC Medical Ltd, Merseyside Innovation Centre, 131 Mount Pleasant, Liverpool L3 5TF

D. Billington School of Biomolecular Sciences, Liverpool John Moores University, Byrom Street, Liverpool L3 3AF

CENTRIFUGATION
RADIOISOTOPES
LIGHT MICROSCOPY
ANIMAL CELL CULTURE
GEL ELECTROPHORESIS: PROTEINS

Forthcoming titles

GENE TECHNOLOGY
PLANT CELL CULTURE
PCR
ANTIBODY TECHNOLOGY
MICROBIAL CULTURE

GEL ELECTROPHORESIS: PROTEINS

M.J. Dunn

Department of Cardiothoracic Surgery, National Heart and Lung Institute, Heart Science Centre, Harefield Hospital, Harefield, Middlesex UB9 6JH, UK

In association with the Biochemical Society

© BIOS Scientific Publishers Limited, 1993

All rights reserved. No part of this book may be reproduced or transmitted, in any form or by any means, without permission.

First published in the United Kingdom 1993 by
BIOS Scientific Publishers Limited,
St Thomas House, Becket Street, Oxford OX1 1SJ

A CIP catalogue for this book is available from the British Library.

ISBN 1 872748 21 X

Typeset by Marksbury Typesetting Ltd, Midsomer Norton, Avon, UK
Printed by The Alden Press Ltd, Oxford, UK

Preface

Electrophoretic methods for the separation and analysis of protein mixtures are used in all disciplines of the biological sciences. The versatility and resolving capacity of these techniques has resulted in their becoming the most popular tool in the armory of the cell and molecular biologist, for the analysis of patterns of gene expression in a wide variety of complex systems. Electrophoresis has also become an indispensable aid to the analytical biochemist. It can be used to characterize protein purity and to monitor the various steps in a protein purification process, whether this is based on traditional methods using biological material as the starting point or on recombinant gene technology. Indeed, the popularity of this group of methods can be judged from the assessment by Dr A.T. Andrews that perhaps as many as three-quarters of all research papers in the whole field of biochemistry make some use of electrophoresis.

The main aim of this book is to provide biological scientists new to the field of electrophoresis with the necessary background and information required to select the electrophoretic method most appropriate to their particular separation problem. The book is, therefore, intended primarily for the advanced undergraduate student, postgraduate scientist, or research technician, but it is hoped that it will also be useful to more experienced scientists wishing to find introductory material to a particular method with which they are unfamiliar.

In the first part of this book the properties of proteins which are exploited in the various techniques of electrophoresis discussed in subsequent sections are outlined. There follows a brief, essentially non-mathematical, treatment of the theory of electrophoresis and the properties of polyacrylamide gels, as an understanding of the principles involved is essential for the proper selection and application of electrophoretic methods to a particular separation problem. The main methods of electrophoresis which allow proteins to be separated on the basis of mobility, size (SDS–PAGE), and charge (IEF) are then described. These one-dimensional methods are able to separate a hundred or so protein components, but the requirement to analyze protein expression in whole tissues and organisms has resulted in the development of two-dimensional electrophoresis methods (2-D PAGE) capable of resolving simultaneously several thousand proteins. The development, application, and variations of this technology are discussed.

The next section of the book deals with the visualization and analysis of the protein patterns resulting from electrophoretic separations. General protein staining and detection methods are described and techniques for the visualization of specific groups of proteins are discussed. Procedures for the extraction of quantitative data from electrophoretic profiles, a relatively easy process for one-dimensional separations but much more complicated in the case of two-dimensional profiles, are described.

Electrophoretic techniques have an almost unrivalled ability to separate the components of complex protein mixtures. However, these methods do not provide directly any information on the identity or functional properties of the separated proteins. In the final section of the book, methods based on Western blotting for the characterization of proteins separated by electrophoresis in terms of their reactivity and affinity with antibodies and other ligands, and their chemical composition are described.

I would like to thank the many scientific colleagues who have allowed me to use examples of their work to illustrate this book, and friends in commercial companies who have provided illustrations of examples of equipment used in the various methods discussed in this volume. I am also grateful to Samantha Crisp, Alan Collins and Annalisa Dunn for their invaluable assistance in the preparation of the illustrations for the book. Finally, I would like to thank Dr John Graham for his continual support and encouragement in the preparation of this volume.

M.J. Dunn

Contents

Abbreviations — xii

PART 1: BASIC PRINCIPLES AND METHODS

1. An Introduction to Electrophoresis — 1

    References — 6

2. Theory of Electrophoresis — 7

    Basic principles — 7
    Electrophoretic mobility — 7
    Effects of pH and ionic strength — 8
    Joule heating — 8
    Zone electrophoresis — 9
    Support media — 9
    Gel-based support media — 10
    Electroendosmosis — 10
    The diversity of electrophoretic techniques — 11
    References — 11

3. Properties of Polyacrylamide Gels — 13

    Gel formation and structure — 13
    Toxicity of acrylamide — 13
    Purity of acrylamide — 15
    Polymerization catalysts — 15
    Pore size — 16
    Alternative cross-linking agents — 17
    Gel configuration — 18
        Cylindrical rod gels — 18
        Slab gels — 19
    Gel concentration — 25
        Homogeneous gels — 25
        Gradient gels — 26
    References — 30

4.	Electrophoresis Under Native Conditions	31
	Continuous buffer systems	31
	Choice of pH and buffer system	31
	Choice of ionic strength	32
	Advantages and disadvantages	33
	Discontinuous buffer systems	33
	Advantages	33
	Mechanism of stacking	34
	Choice of discontinuous buffer system	35
	Estimation of molecular mass of native proteins	36
	Ferguson plot	36
	Pore gradient electrophoresis	37
	Transverse gradient gel electrophoresis	38
	References	39
5.	Electrophoresis in the Presence of Additives	41
	Disulfide bond cleaving agents	41
	Urea	41
	Transverse gradients of urea	42
	Detergents	43
	Non-ionic detergents	45
	Zwitterionic detergents	47
	Anionic detergents	48
	Cationic detergents	48
	References	49

PART 2: TECHNIQUES AND APPLICATIONS

6.	Sodium Dodecyl Sulfate–Polyacrylamide Gel Electrophoresis (SDS–PAGE)	51
	Basic principles	51
	Choice of gel concentration	51
	Buffer systems	52
	Continuous buffer systems	52
	Discontinuous buffer systems	52
	Sample preparation	53
	Estimation of molecular mass by SDS–PAGE	54
	Homogeneous gels	54
	Gradient gels	55
	Limitations of SDS–PAGE	57
	SDS–PAGE in non-reducing conditions	58
	Peptide mapping by SDS–PAGE	59
	The need for peptide mapping	59
	In situ peptide mapping	61
	Primary gel system	61

	Protein cleavage	62
	Secondary gel system	62
	References	64
7.	**Isoelectric Focusing**	**65**
	Background	65
	IEF using synthetic carrier ampholytes	65
	Basic principles	65
	Gel concentration	67
	Gel composition	67
	Choice of ampholytes	67
	Gel preparation and apparatus	69
	Running conditions	70
	Estimation of pH gradient	71
	Limitations of IEF using synthetic carrier ampholytes	73
	IEF using immobilized pH gradients	74
	Basic principles	74
	pH gradients	75
	Problems of Immobiline reagents	77
	Gel preparation	77
	Sample application	79
	Running conditions	80
	Estimation of pH gradient	81
	Titration curves	81
	Characterization of genetic mutants	81
	Analysis of macromolecular interactions	82
	Estimation of pK values	83
	References	85
8.	**Two-Dimensional Gel Electrophoresis**	**87**
	Background	87
	Sample preparation	88
	Body fluids	89
	Solid tissue samples	90
	Circulating and cultured cells	91
	Plant tissues	91
	Sample solubilization	91
	The first dimension	93
	IEF using synthetic carrier ampholytes	93
	Non-equilibrium pH gradient electrophoresis	94
	Immobilized pH gradients	95
	Gel recovery	99
	Estimation of pH gradients	100
	Equilibration between dimensions	100
	Transfer between dimensions	101
	The second dimension	102

	Molecular mass standards	102
	Gel size, resolution and reproducibility	103
	Two-dimensional PAGE under native conditions	104
	Two-dimensional PAGE under denaturing conditions	105
	Ribosomal proteins	105
	Nuclear proteins	106
	Membrane proteins	109
	References	109
9.	**Detection Methods**	**111**
	Gel recovery	111
	Fixation	111
	General protein stains	112
	Coomassie brilliant blue	113
	Fluorescent staining methods	115
	Silver staining	115
	Fixation	116
	Enhancement	117
	Staining and development	117
	Radioactive detection methods	119
	Radiolabeling methods	119
	Gel drying	120
	Autoradiography	121
	Fluorography	121
	Intensification screens	122
	Dual isotope methods	122
	Electronic detection methods	123
	Gel fractionation and counting	123
	Detection of glycoproteins	124
	Detection of phosphoproteins	125
	Detection of lipoproteins	125
	Detection of enzymes	125
	References	126
10.	**Quantitative Analysis**	**129**
	Background	129
	Gel imaging	129
	Densitometers	129
	Laser densitometers	130
	TV cameras	130
	CCD array scanners	131
	Radioisotope imagers	132
	Photostimulable phosphor-imaging systems	133
	Analysis of one-dimensional gels	135
	Analysis of two-dimensional gels	136
	References	138

11. **Western Blotting**	**139**
Background	139
Transfer methods	141
Blotting matrices	144
Transfer buffers	145
General protein staining	146
Protein standards	147
Blocking	147
Specific detection	148
Quantitation	152
References	152
12. **Chemical Characterization of Proteins Separated by Electrophoresis**	**153**
Background	153
Recovery of proteins from gels	153
Elution by diffusion	154
Electrophoretic elution	154
Protein characterization on Western blots	155
Amino acid analysis	155
N-terminal protein sequence analysis	156
Internal protein sequence analysis	159
References	161
Appendices	**163**
Appendix A: Glossary	163
Appendix B: Suppliers	166
Appendix C: Further reading	169
Index	171

Abbreviations

AMPPD	disodium 3-(4-methoxyspiro[1,2-dioxetane-3-2'-tricyclo- [3.3.3.1^{3,7}]decan]4-yl)phenyl phosphate
ANS	1-aniline-8-naphthalene sulfonate
APAP	alkaline phosphatase–antialkaline phosphatase
BAC	N,N'-*bis*-acrylylcystamine
BCIP	5-bromo-4-chloroindoxyl phosphate
Bis -	N,N'-methylene-*bis*-acrylamide
BNPS-skatole	2-(2'-nitrophenylsulfenyl)-3-methyl-3-bromoindolenine
BSS	balanced salt solution
CAPS	3-(cyclohexylamino)-1-propanesulfonic acid
CBB	Coomassie brilliant blue
CCD	charge-coupled device
CHAPS	3-[(cholamidopropyl)dimethylammonio]-1-propane sulfonate
CK	creatine kinase
CSF	cerebrospinal fluid
CTAB	cetyltrimethylammonium bromide
CZE	continuous zone electrophoresis
Da	dalton (atomic mass unit)
DAB	diaminobenzidine
DATD	N,N'-diallyltartardiamide
DHEBA	N,N'-(1,2-dihydroxyethylene)*bis*-acrylamide
DMAPMA	3-dimethylaminopropyl methacrylamide
DMSO	dimethylsulfoxide
DNA	deoxyribonucleic acid
DNase	deoxyribonuclease
DOC	sodium deoxycholate
DTT	dithiothreitol
E. coli	*Escherichia coli*
ECL	enhanced chemiluminescence
EDIA	ethylenediacrylate
EDTA	ethylenediaminetetraacetic acid
EEO	electroendosmosis

FITC	fluorescein isothiocyanate
HPLC	high-performance liquid chromatography
IEF	isoelectric focusing
IgG	immunoglobulin G
IPG	immobilized pH gradient
K_d	dissociation constant
K_r	retardation coefficient
M_r	relative molecular mass
MAPTAC	methacrylamidopropyltrimethylammonium chloride
MDPF	2-methoxy-2,4-diphenyl-3(2H)-furanone
mRNA	messenger RNA
MZE	multiphasic zone electrophoresis
NBT	nitroblue tetrazolium
NEPHGE	non-equilibrium pH gradient electrophoresis
OD	optical density
OPA	o-phthalaldehyde
PAGE	polyacrylamide gel electrophoresis
PAP	peroxidase–antiperoxidase
PBS	phosphate-buffered saline
pI	isoelectric point
PITC	phenylisothiocyanate
PMSF	phenylmethanesulfonyl fluoride
PPO	2,5-diphenyloxazole
PTH	phenylthiohydantoin
PVDF	polyvinylidene difluoride
R_f	relative mobility
RNA	ribonucleic acid
RNase	ribonuclease
RP-HPLC	reverse-phase high-performance liquid chromatography
SDS	sodium dodecyl sulfate
TACT	N,N',N''-triallylcitric triamide
TCA	trichloroacetic acid
TEMED	N,N,N',N'-tetramethylethylenediamine
TFA	trifluoroacetic acid
UV	ultraviolet
Vh	volthours
1-D	one-dimensional
2-D	two-dimensional

1 An Introduction to Electrophoresis

The genetic information in an organism is contained within the nucleus where it is arranged into genes encoded by the DNA which constitutes the chromosomes. The regulation of the differential activity of these genes (i.e. the genotype) is of primary importance in determining the

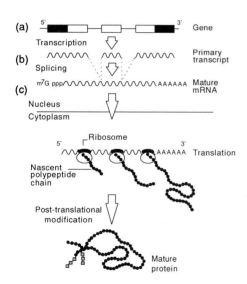

FIGURE 1.1: The overall process of gene expression is shown in three steps. (**a**) RNA polymerase, which binds to the 5' end of the gene, moves along the DNA transcribing the linear sequence of DNA into a corresponding sequence of RNA. Transcription always occurs in the 5' to 3' direction. (**b**) During splicing, unwanted intron sequences are removed from the RNA transcript, and the exon sequences are joined. (**c**) The mature mRNA is transported into the cytoplasm of the cell where it is bound by ribosomes and the linear sequence of RNA is translated into protein. Translation occurs in the 5' to 3' direction, and starts with the amino (NH_2) terminus of the nascent polypeptide. A variety of post-translational modifications can be made to a protein, for example, the covalent linking of oligosaccharide chains (represented by the squares) to produce a glycoprotein. Adapted from reference [6] with permission from Current Science.

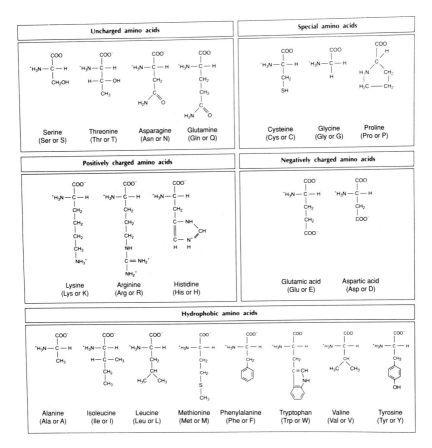

FIGURE 1.2: *The structure of the 20 commonly occurring amino acids. In each amino acid the α-carbon atom is bonded to an amino group (or imino group in the case of proline), a carboxyl group, a hydrogen atom, and a side chain responsible for its individuality. Reproduced from reference [7] with permission from Current Science.*

actions and properties (i.e. the phenotype) of individual cells within an organism under various conditions, such as during development and in response to disease processes and experimental conditions.

During gene expression the information encoded within the genome is converted by a process of transcription to a corresponding primary RNA transcript (*Figure 1.1*). This is subsequently processed within the nucleus to form mature messenger RNA (mRNA). This mRNA is transported to the cytoplasm where it is bound to ribosomes, to act as a template for translation into the corresponding protein (*Figure 1.1*). The resulting proteins are the primary working molecules within the cell and a cell's phenotype is determined by the proteins which it expresses. The diversity of the functional properties of proteins is truly remarkable, for example: (i) acting as enzymes to catalyze the majority

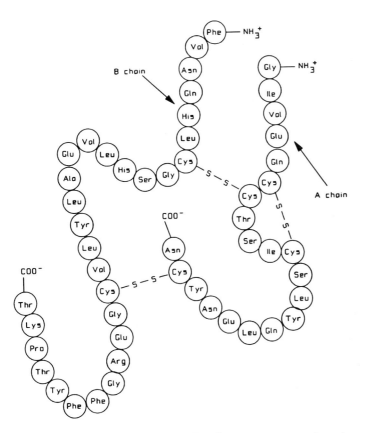

FIGURE 1.3: *Human insulin. The linear sequence of amino acids in each polypeptide chain (A and B) determines the primary structure. In addition, two disulfide bridges connect residues of the A and B chains and one bridge connects two residues of the A chain.*

of cellular chemical reactions, (ii) providing mechanical support, (iii) controlling membrane permeability, transport, and storage processes, (iv) acting as hormones, receptors, and signal transducers, (v) providing immune protection, (vi) being responsible for co-ordinated motion, and (vii) controlling gene expression.

The primary building blocks of proteins are the 20 different naturally occurring amino acids (*Figure 1.2*). The amino acids are arranged in a linear array, each being connected by a peptide bond formed between the amino group of one amino acid and the carboxyl group of another, to form a polypeptide (*Figure 1.3*). This linear sequence of amino acids is specifically determined by the corresponding sequence of nucleotides in the DNA constituting the genes and determines the primary structure of a protein. The secondary structure of a protein is defined by the folding of regions of the polypeptide chain into regular structures, the most important structural elements being the α-helix, the β-pleated sheet and the β-turn (*Figure 1.4*). The tertiary structure of a protein is determined by the folding of regions between α-helices and β-pleated sheets and by the condensation of these features into compact shapes known as domains (*Figure 1.5*). The tertiary structure is maintained by both covalent (e.g. disulfide bonds) and non-covalent bonds. An additional level of structural organization, termed quaternary structure,

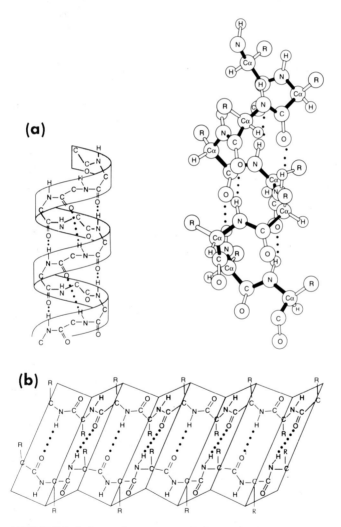

FIGURE 1.4: *Secondary structural elements of proteins. (a) α-helix. The ribbon representation emphasizes the helical form. The ball-and-stick representation indicates the role of individual atoms and shows the amino acid side chains (R) that protrude from the helix. (b) β-pleated sheet formed by hydrogen bonding between polypeptide chains. The planarity of the peptide bond forces the structure to be pleated. Reproduced from reference [7] with permission from Current Science.*

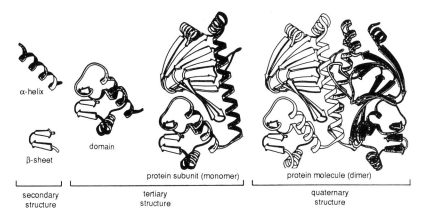

FIGURE 1.5: *The three-dimensional structure of a protein can be described in terms of different levels of folding, each of which is constructed from the preceding one in hierarchical fashion. Reproduced from reference [1] with permission from Garland Publishing.*

exists for the many proteins which contain more than one polypeptide and refers to the spatial arrangement of the constituent polypeptides or subunits (*Figure 1.5*).

Those readers wishing to explore in further detail the whole area of molecular cell biology are referred to some of the excellent recent textbooks on this subject [1–4].

The side chains of four of the amino acids are highly ionized and are, therefore, charged at the near neutral pH values which exist in most biological systems *in vivo*. Glutamic acid and aspartic acid are negatively charged, while lysine and arginine are positively charged (*Figure 1.2*). Histidine is also positively charged, but is only weakly ionized at neutral pH. Proteins are, therefore, zwitterions and the net charge of a protein at a particular pH is determined by its content of acidic and basic amino acids and by their degree of ionization at that pH. It is clear, then, that at nearly all pH values proteins carry a net charge which is either negative or positive, depending on the particular pH value. In fact, at physiological pH, the majority of both animal and plant proteins are negatively charged, but nevertheless, there are major groups of proteins which are positively charged under these conditions. For every protein there is a particular pH, known as the isoelectric point (pI), at which the negative and positive charges on the protein are in balance, so that the protein has a net charge of zero.

These electrical properties of proteins are exploited in the various techniques of electrophoresis described in this book, to achieve the effective separation and resolution of mixtures of proteins required for their analysis and further characterization. The versatility and resolving capacity of these techniques has resulted in their becoming the most popular tool in the armory of the cell and molecular biologist, for the analysis of patterns of gene expression in a wide variety of complex systems. Electrophoresis has also become an indispensable aid to the analytical biochemist. It can be used to characterize protein purity and to monitor the various steps in a protein purification process, whether this is based on traditional methods using biological material as the starting point or on recombinant gene technology. Indeed, the popularity of this group of techniques can be judged from the assessment of Andrews [5] that "perhaps as many as three-quarters of all research papers in the whole field of biochemistry make some use of electrophoresis".

References

1. Alberts, B., Bray, D., Lewis, J., Raff, M., Roberts, K. and Watson, J.D. (1989) *Molecular Biology of the Cell,* 2nd Edn. Garland, New York.
2. Lewin, B. (1990) *Genes IV*. Oxford University Press, Oxford.
3. Darnell, J., Lodish, H. and Baltimore, D. (1990) *Molecular Cell Biology*. Scientific American Books, New York.
4. Watson, J.D., Gilman, M., Witkowski, J. and Zoller, M. (1992) *Recombinant DNA,* 2nd Edn. Scientific American Books, New York.
5. Andrews, A.T. (1986) *Electrophoresis: Theory, Techniques and Biochemical and Clinical Applications*. Clarendon Press, Oxford.
6. Barton, P.J.R. (1991) in *Ann. Cardiac Surgery* (M. Yacoub and J. Pepper, eds). Current Science, London, p. 3–12.
7. Dunn, M.J. (1992) in *Ann. Cardiac Surgery* (M. Yacoub and J. Pepper, eds). Current Science, London, p. 1–8.

2 Theory of Electrophoresis

2.1 Basic principles

This introduction to electrophoresis has been written without resort to a mathematical treatment of the physical chemistry of the subject. However, extensive work has been carried out in this area and those interested in exploring in more detail the theory underlying electrophoresis are referred to a recent monograph [1].

Electrophoresis is defined as the migration of a charged particle in an electric field. Under conditions of constant velocity, the driving force on a particle is the product of the charge on the particle and the applied field strength. This force is counteracted by the frictional resistance of the separation medium, which is proportional to its shear velocity. While Stokes' law is obeyed in free solution, the situation is complicated if a gel medium is used, so that frictional resistance also depends on additional factors such as gel density and particle size.

2.2 Electrophoretic mobility

An important concept in electrophoresis is electrophoretic mobility, which is defined as the velocity of the particle per unit field strength. It is clear, then, that the choice of applied voltage and separation path length, which determine the field strength, together with the time of the run are important parameters in optimizing an electrophoretic separation.

2.3 Effects of pH and ionic strength

As we have seen in Chapter 1, proteins possess both negatively and positively charged groups as part of their primary structure, so that they act as zwitterions. Because the net charge on a protein is dependent upon the pH of the medium, pH will exert a profound influence on protein mobility during electrophoresis. If the operative pH is the same as the pI of the protein, it will not migrate during electrophoresis. At pH values below its pI, the protein will move towards the cathode, while at pH values above its pI, it will migrate towards the anode.

The ionic strength of the separation medium also exerts a major influence on electrophoretic mobility. Buffers of low ionic strength permit higher rates of migration than do those of higher ionic strength, while the latter generally result in sharper zones of separation. As an approximation, mobility is inversely proportional to the square root of the ionic strength [2]. Thus, choice of buffer ionic strength is an important parameter in determining the time and resolution of an electrophoretic separation.

2.4 Joule heating

During every electrophoretic separation, electrical energy is transformed into heat, termed Joule heating. This can result in severe deleterious effects such as enzyme denaturation, increased diffusion of protein molecules (which degrades resolution), and even damage to the electrophoretic equipment itself. The limitation and removal of this generated heat is, therefore, a major consideration in the implementation of a particular electrophoretic separation and in the design of the equipment used. The choice of buffer strength is crucial here, as the higher its ionic strength, the greater its conductivity, and the greater the amount of heat generated. Depending on which buffer system is selected, electrical resistance can either remain constant, increase, or decrease as electrophoresis proceeds. In the case of a homogeneous buffer system (see Section 4.1), resistance remains constant during electrophoresis so that Joule heating can be controlled simply by power input. However, in the case of a discontinuous (multiphasic) buffer system (see Section 4.2), the electrical resistance of the gel increases as the moving boundary migrates through the gel, due to a decrease in conductance. Thus, if a constant current is applied, voltage and Joule heating will increase with time. In contrast, at a constant applied voltage, the current and hence Joule heating will decrease with time, but at the expense of an increased separation and generally reduced resolution.

It is desirable to use power supplies which can be regulated to provide a constant output during electrophoresis. Nearly all commercially available power supplies (see Appendix B) provide for use of either constant current, to be used in systems with decreasing or constant resistance, or constant voltage, to be used where resistance increases during the separation. Some more expensive power packs have the capability to monitor continuously voltage and current, thereby performing electrophoresis at constant power. This approach provides constant heat generation during electrophoresis, but other factors such as choice of buffer, applied voltage, apparatus design, and cooling are critical for the minimization and removal of that heat. In any case, reproducibly controlled separations can only be achieved by carrying out electrophoresis at a constant temperature.

2.5 Zone electrophoresis

As proteins are charged at any pH other than their pI, they will migrate in an electric field at a rate which is dependent on their charge density. Charge density is defined as the ratio of charge to mass. The higher this ratio, the faster the molecule will migrate. Thus, if an electric field is applied to a solution of proteins, the different molecules will migrate at different rates (dependent on the magnitude of their charge density) towards either the anode or the cathode (dependent on whether their net charge is negative or positive). Little or no separation of the proteins is possible if the proteins are present throughout the separation medium. However, if the sample is initially present as a narrow zone, proteins of different mobilities will travel as discrete zones and thus separate during electrophoresis. This approach is known as zone electrophoresis.

2.6 Support media

It is possible to carry out zone electrophoresis in free solution. Indeed, this method was used in many of the early electrophoretic experiments, but this approach has severe disadvantages. In particular, Joule heating is, as we have seen, an inevitable by-product of electrophoresis and this will result in convective effects which disrupt the separated protein zones. In addition, diffusion effects will operate to broaden the protein zones and result in degradation of the separation. These deleterious effects can be minimized by performing the electrophoretic separation in a support medium which inhibits convection. The earliest such support to be used was porous paper, but this was soon supplanted by other supports such as cellulose acetate and gel-based media such as starch, agarose, and polyacrylamide.

2.7 Gel-based support media

A major advance in electrophoresis occurred with the advent of polyacrylamide gels [3]. Polyacrylamide gels are non-ionic polymers which are chemically inert, stable over a wide range of pH, temperature and ionic strength, and are transparent. Moreover, polyacrylamide gels can be produced with a wide range of pore sizes, optimized for the separation of proteins of different size ranges. Because of these advantages, together with the fact that very high resolution protein separations can be obtained, polyacrylamide gels have become the support medium of choice for zone electrophoresis of proteins.

Very large molecules cannot be separated satisfactorily using polyacrylamide gels, so that agarose gels are used for the analysis of large molecules or complexes (e.g. nucleoproteins, nucleic acids). Agarose is also usually employed in immunoelectrophoretic techniques [4] and starch gels are widely used for the analysis of isoenzymes [5].

In the case of gel-based media it is important to understand that their relatively small pore size, approximately of the same order as the size of protein molecules, will result in a molecular sieving effect during electrophoresis, so that the resulting separation will depend on both the charge density and size of the proteins being analyzed.

2.8 Electroendosmosis

Additional factors resulting from the nature of the support medium can also influence protein mobility and resolution during electrophoresis. In particular, charged groups present on the support can cause problems. First, this can result in an ion-exchange effect of the support with the proteins. Perhaps more importantly, these fixed charged groups result in electroendosmosis. In the (usual) case of negatively charged groups present on the matrix, these are attracted towards the anode during electrophoresis. This is, of course, not possible as the charged groups are immobilized on the support, so the effect is compensated for by a migration of H^+ ions, as hydrated protons (H_3O^+), towards the cathode. This results in a flow of water towards the cathode, which is often deleterious to the separation. This effect can be pronounced using paper supports or certain preparations of agarose, but is fortunately small in the case of polyacrylamide gels.

2.9 The diversity of electrophoretic techniques

The basic concept of electrophoresis is simple, but it should by now be clear that there are a wide variety of factors which can influence and modify an electrophoretic separation process. Indeed, these various factors have been exploited to generate a diversity of electrophoretic techniques able to separate proteins according to size, mobility, net charge, and even hydrophobicity. These various techniques will form the basis of the subsequent chapters of this book.

References

1. Mosher, R.A., Saville, D.A. and Thormann, W. (1992) *The Dynamics of Electrophoresis*. VCH, Weinheim.
2. Allen, R.C., Saravis, C.A. and Maurer, H.R. (1984) *Gel Electrophoresis and Isoelectric Focusing of Proteins*. Walter de Gruyter, Berlin.
3. Raymond, S. and Weintraub, L.S. (1959) *Science*, **130**, 711–721.
4. Bog-Hansen, T.C. (1990) in *Gel Electrophoresis of Proteins: a Practical Approach* (B.D. Hames and D. Rickwood, eds). IRL Press, Oxford, p. 273–300.
5. Siciliano, M.J. and Shaw, C.R. (1976) in *Chromatographic and Electrophoretic Techniques*, Vol. II. Heinemann, London, p. 185–209.

3 Properties of Polyacrylamide Gels

3.1 Gel formation and structure

Polyacrylamide gel is formed by the polymerization of monomers of acrylamide (*Figure 3.1a*) with monomers of a suitable bifunctional cross-linking agent. The best and most commonly used cross-linking agent for the majority of electrophoretic applications is N,N'-methylene-*bis*-acrylamide (*Figure 3.1b*), referred to as Bis for short. A three-dimensional network is formed by the cross-linking of randomly growing linear polyacrylamide chains by a mechanism of vinyl polymerization (*Figure 3.1c*). The concentration of acrylamide used determines the average length of the linear chains of the polymer, while the concentration of Bis determines the extent of cross-linking. Both of these parameters are, therefore, important in determining the physical properties of the gel, including pore size, elasticity, density, and mechanical strength. An important property of polyacrylamide gels is that they are transparent and this is especially important for the satisfactory visualization of the separated protein components after electrophoresis.

For practical purposes, a polyacrylamide gel is normally considered to be composed of a three-dimensional lattice of long polyacrylamide chains, tied at intervals by 'knots' of cross-linking agent. However, the actual structure of polyacrylamide gels is far from clear and is still a controversial subject. Any reader interested in exploring this topic in more detail is referred to references [1–3].

3.2 Toxicity of acrylamide

It is important to be aware that acrylamide is an accumulative neurotoxin and that handling can easily generate airborne particulates which may be inhaled. For this reason, acrylamide powder must

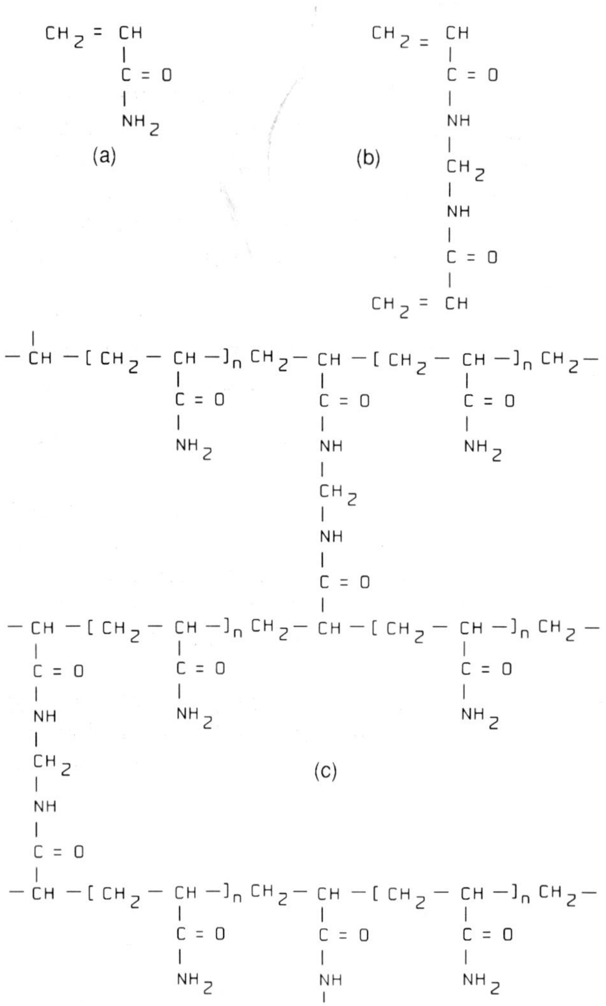

14 GEL ELECTROPHORESIS: PROTEINS

(a)
$$CH_2 = CH$$
$$|$$
$$C = O$$
$$|$$
$$NH_2$$

(b)
$$CH_2 = CH$$
$$|$$
$$C = O$$
$$|$$
$$NH$$
$$|$$
$$CH_2$$
$$|$$
$$NH$$
$$|$$
$$C = O$$
$$|$$
$$CH_2 = CH$$

(c)

$$-CH-[CH_2-CH-]_n CH_2-CH-[CH_2-CH-]_n CH_2-$$

(with repeating side chains: $C=O$, NH_2 / NH–CH_2–NH–$C=O$ cross-links / NH_2, etc.)

FIGURE 3.1: *The chemical structures of (a) acrylamide and (b) the cross-linking agent, N,N'-methylene-bis-acrylamide, which co-polymerize to form (c) the polyacrylamide gel.*

always be dispensed in a fume cupboard; the operator should wear a face mask, safety goggles, and protective gloves. The latter must also be worn when handling acrylamide solutions. If you are unsure about the proper procedures you should consult your Safety Officer. The potential toxic hazard of acrylamide has resulted in several companies producing ready-made forms of the reagent suitable for electrophoresis. These are either pre-weighed powders or solutions, with the acrylamide and Bis being either separate or mixed (see Appendix B).

3.3 Purity of acrylamide

The chemical purity of the acrylamide and Bis used should also be considered. General laboratory grade reagents are likely to be impure, containing breakdown products such as acrylic acid, polyacrylic acid, ammonia, and β',β'',β'''-nitrilo-*tris*-proprionamide. If used, such reagents must be recrystallized, using chloroform for acrylamide and acetone in the case of Bis, prior to use for electrophoresis. The toxic nature of acrylamide renders such a procedure undesirable and purchase of commercially purified supplies of monomer, which have been quality-controlled for electrophoresis, from one of the specialist suppliers (see Appendix B) is strongly recommended.

It is important to realize that all stocks of acrylamide, whatever their initial purity and irrespective of whether they are in powder or liquid form, will break down on prolonged storage. The formation of charged breakdown products, in particular acrylic acid, is deleterious to most electrophoretic techniques and can be disastrous for isoelectric focusing (see Chapter 7). Certain manufacturers claim to have added inhibitors to their reagents to prolong their shelf life. However, it is relatively easy to remove charged breakdown products from an acrylamide solution prior to gel formation by mixing 100 ml with 1 g of a suitable ion exchange resin (e.g. Amberlite MB-1 mono-bed resin) for at least 1 h. The resin can easily be removed by filtration and the resulting solution used for gel polymerization.

3.4 Polymerization catalysts

Gel polymerization (*Figure 3.1c*) is usually initiated with ammonium persulfate and the reaction is accelerated by the addition of the catalyst, N,N,N',N'-tetramethylethylenediamine (TEMED). This reaction is more efficient than the alternative method of u.v.-activated polymerization with riboflavin, but careful control of polymerization conditions is essential if consistent results are to be obtained. Riboflavin can be used to advantage as a catalyst when proteins which are particularly sensitive to persulfate ions are to be separated, otherwise an extensive period of pre-electrophoresis prior to sample application would be required if persulfate was used. The other advantage of riboflavin is that polymerization is not initiated until the gel solution is exposed to u.v. illumination. It should be noted that dissolved oxygen will inhibit the process of polymerization, so that gel mixtures should be thoroughly degassed using a rotary vacuum pump or water suction pump.

The optimum temperature for gel polymerization is in the range 25–30°C, when the reaction generally occurs within a few minutes. However, it is usual to allow at least 2 h (or overnight) for complete polymerization to occur before gels are used for electrophoresis. The use of high temperatures (>50°C), while it does accelerate the speed of polymerization, is not recommended as it can result in the formation of short polymer chains and inelastic gels. Polymerization kinetics are not favorable at low temperatures (0–4°C), but precooling of acrylamide solutions is often a good idea if it is necessary to prolong the polymerization time (e.g. when casting gradient gels, see Section 3.8.2).

3.5 Pore size

The effective size of the pores within a polyacrylamide gel matrix is dependent both on the total concentration of acrylamide monomers plus cross-linking agent, and on the concentration of the cross-linking agent alone. Gel composition is defined by two parameters, %T and %C, where:

%T = acrylamide monomer + cross-linking agent as % (w/v),

and

%C = percentage (by weight) of the total monomer which is cross-linking agent.

The pore size of polyacrylamide gels can be progressively increased by reducing %T at a fixed %C, but very dilute gels (< 2.5%T) are mechanically unstable, thereby limiting the maximum effective pore size to about 80 nm; these are suitable for resolving molecules up to a relative molecular mass of (M_r) 10^6. It is possible to prepare polyacrylamide gels at very high values of %T (> 30%T); such gels have very small effective pores which are able to restrict the passage of molecules with M_r as low as 2×10^3. Therefore, the choice of acrylamide concentration (i.e. %T) is critical for optimizing an electrophoretic protocol for any particular protein sample requiring analysis and this is discussed in more detail in Section 3.8.

As the proportion of cross-linking agent is increased at a fixed %T, the effective pore size also decreases. However, the pore size reaches a minimum at a particular %C which is dependent on the value of %T. Further increase in the value of %C results in an increase in the effective pore size of the gel and this is thought to be due to the formation of bead-like structures within the gel, rather than a three-dimensional network [2]. Stable gels of high pore size (< 250 nm) can be made in this way, but at concentrations of Bis in excess of 30%C the gels become unacceptably opaque, hydrophobic, and prone to collapse.

3.6 Alternative cross-linking agents

The problems associated with using Bis at high values of %C can be partially overcome by using the alternative cross-linking agent, N,N'-(1,2-dihydroxyethylene)*bis*-acrylamide (DHEBA) (*Figure 3.2e*). This can be used at values of up to 50%T, giving gels with effective pore sizes in the region of 500–600 nm.

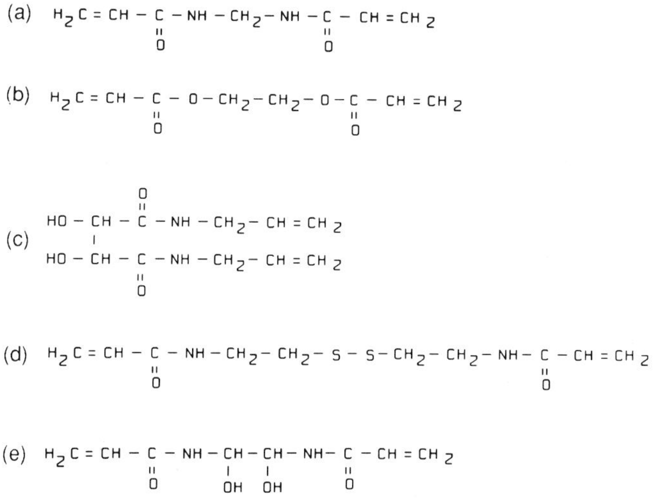

(a) $H_2C = CH - C(=O) - NH - CH_2 - NH - C(=O) - CH = CH_2$

(b) $H_2C = CH - C(=O) - O - CH_2 - CH_2 - O - C(=O) - CH = CH_2$

(c) $HO - CH - C(=O) - NH - CH_2 - CH = CH_2$
    $HO - CH - C(=O) - NH - CH_2 - CH = CH_2$

(d) $H_2C = CH - C(=O) - NH - CH_2 - CH_2 - S - S - CH_2 - CH_2 - NH - C(=O) - CH = CH_2$

(e) $H_2C = CH - C(=O) - NH - CH(OH) - CH(OH) - NH - C(=O) - CH = CH_2$

FIGURE 3.2: *Alternative cross-linking agents used for the formation of polyacrylamide gels. (**a**) N,N'-methylene-bis-acrylamide (Bis); (**b**) ethylenediacrylate (EDIA); (**c**) N'N'-diallyltartardiamide (DATD); (**d**) N,N'-bis-acrylylcystamine (BAC); and (**e**) N,N'-(1,2-dihydroxyethylene)-bis-acrylamide (DHEBA).*

The properties of alternative cross-linking agents can also be exploited to produce reversible gels, so that they can be solubilized after the electrophoretic separation is complete. For example, the 1,2-diol structure of DHEBA renders gels susceptible to cleavage by oxidation with periodic acid. Some of these alternative cross-linkers are shown in *Figure 3.2* and this topic is discussed in detail in references [2] and [3]. It is recommended that the use of the agents, N,N'-diallyltartardiamide

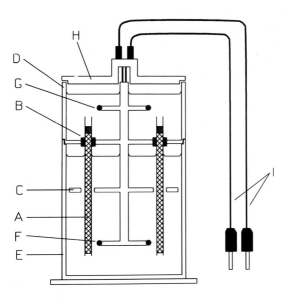

FIGURE 3.3: *Diagram of typical apparatus for running cylindrical polyacrylamide gels. A, Gels contained in glass tubes; B, rubber grommets; C, tube locating ring; D, upper chamber; E, lower chamber; F, lower electrode (usually anode); G, upper electrode (usually cathode); H, lid; I, connections to power pack.*

(DATD) and N,N',N''-triallycitric triamide (TACT), should be avoided as they have been found to be inhibitors of polymerization.

3.7 Gel configuration

3.7.1 Cylindrical rod gels

The original procedures for performing zone electrophoresis in polyacrylamide gels were developed using cylindrical gels polymerized in glass tubes. The tubes which are used normally have an internal diameter (i.d.) of 3–5 mm and gels varying in length from about 5 to 15 cm can be prepared. In order to obtain reproducible separations, it is essential that the tubes all have the same i.d. and the gels are all cast to the same precise length. The design of an apparatus suitable for running cylindrical polyacrylamide gels is shown diagrammatically in *Figure 3.3* and a typical commercial apparatus is shown in *Figure 3.4*.

To prepare gels in glass tubes, the bottom ends of the tubes must first be sealed with small caps or bungs, or by wrapping the ends with wax film (Parafilm). The tubes are then placed in a rack to ensure that they are vertical and the appropriate volume of gel solution to prepare gels

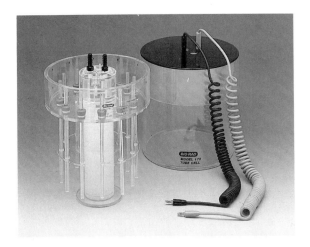

FIGURE 3.4: *Typical commercial apparatus for running cylindrical polyacrylamide gels. (Courtesy of Bio-Rad Laboratories Ltd.)*

of the required length is introduced using a syringe or pipet. The gels are then overlaid with water-saturated butan-2-ol to exclude oxygen, which would inhibit gel polymerization; this also produces a flat gel surface which is essential if straight protein zones are to be obtained during electrophoresis. After polymerization is complete, the tops of the gels should be flushed with water to remove the butan-2-ol, which can otherwise induce protein precipitation during electrophoresis. The tubes are then mounted into the apparatus and the electrode chambers filled with the appropriate buffer. The samples can then be applied using a microsyringe and electrophoresis performed for the appropriate time.

Although cylindrical rod gels have largely been superseded by slab gels (see Section 3.7.2), they are still used under some circumstances. In particular they are widely used for the first dimension in two-dimensional electrophoretic techniques (see Section 8.4). They are also useful when the gel is to be sliced into small transverse segments after electrophoresis, in order either to measure the amount of radioactive protein in each slice or to elute and assay the separated proteins for biological assay.

3.7.2 Slab gels

Flat slab gels are currently the method of choice for electrophoresis in one dimension, as many samples can be electrophoresed simultaneously on the same gel under identical conditions, thereby increasing the reproducibility and comparability of the separation patterns (*Figure 3.5*).

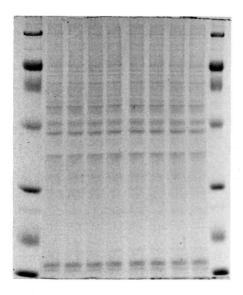

FIGURE 3.5: *Separation of the total proteins of human endothelial cells by slab gel electrophoresis. The lanes at either side are molecular weight markers.*

Slab gels are normally cast in a slab gel cassette, which in its simplest form consists of two glass plates separated by removable spacers (*Figure 3.6*). The thickness of the gel is determined by the thickness of these spacers and for analytical purposes is normally in the range 0.5–1.5 mm. Thinner gels are generally preferred since they are easier to cool during electrophoresis. However, thicker gels of up to 6 mm can be used for preparative applications. The cassette is then clamped together and assembled into some form of gel casting stand to ensure a water-tight assembly. The gel is prepared by introducing the required volume of degassed gel solution into the cassette and the wells to contain the individual samples are formed by insertion of a plastic comb around which the gel polymerizes. The dimensions of the 'teeth' of this comb determine the number and size of the resulting wells and the volume of sample which can subsequently be applied. Most users of electrophoretic techniques still prefer to prepare their own polyacrylamide

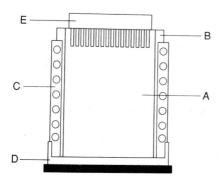

FIGURE 3.6: *A cassette for casting polyacrylamide slab gels. A, Glass plates; B, spacer; C, clamps; D, stand sealing bottom of cassette; E, plastic comb to form sample wells.*

gels due to the versatility that this provides in terms of gel size, composition (%T and %C), additives, and buffer system. However, a number of companies can now supply a range of ready-made gels suitable for use with both vertical and horizontal slab gel apparatus (see Appendix B).

a. Vertical slab gels. The most popular technique for electrophoresis in slab gels is to use an apparatus which holds the slab in a vertical position between the two electrode reservoirs. Suitable commercial equipment is available from a wide range of suppliers (see Appendix B), but can be broadly categorized into two types. The first type is for use with standard format gels, typically 14 cm wide by 16 cm long (*Figure 3.7*). Such systems are very versatile and are the most commonly used for protein electrophoresis. The second type of apparatus uses a miniaturized gel format, with the gels typically being about 7 cm in length (*Figure 3.8*). The main advantage of miniaturized systems is that they provide short run times compared with standard format gels, which has made mini-gel techniques very popular for a variety of different rapid screening techniques. Whichever type of apparatus is used, it is essential that efficient cooling is provided if good separations are to be achieved without distortion or 'smiling' of the protein zones.

b. Horizontal slab gels. An alternative procedure is to carry out electrophoresis using horizontal slab gels. This technology has been

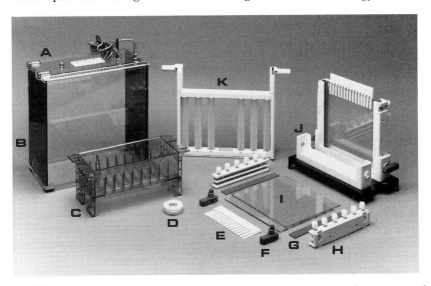

FIGURE 3.7: *Typical commercial vertical slab gel electrophoresis apparatus for use with standard format (14 × 16 cm) polyacrylamide gels. A, Lid with electrode connections; B, lower chamber; C, upper chamber; D, grease for spacers; E, plastic comb for forming sample wells; F, cam to attach cassette to casting stand; G, spacers; H, clamps; I, glass plates; J, casting stand; K, cooling coil. (Courtesy of Hoefer Scientific Instruments.)*

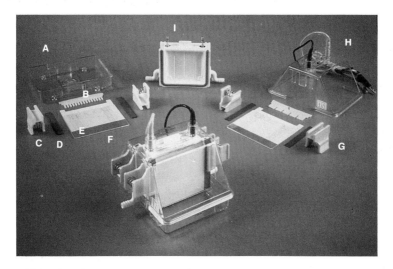

FIGURE 3.8: *Typical miniaturized commercial vertical slab gel electrophoresis apparatus for use with small format (7 cm) polyacrylamide gels. A, Lower chamber; B, plastic comb for forming sample wells; C, clamps; D, spacers; E, ceramic back plate; F, glass plate; G, clamps; H, lid with electrode connections; I, cooling plate. (Courtesy of Hoefer Scientific Instruments.)*

pioneered by Dr Angelika Görg and her colleagues in Munich [4, 5] and commercialized largely by Pharmacia LKB Biotechnology. In this approach the gel is placed on the horizontal cooling platten of a flat-bed electrophoresis apparatus (*Figure 3.9*). The design of a typical apparatus is shown diagrammatically in *Figure 3.10*. Temperature control during electrophoresis is dependent on efficient cooling of the gel by the cooling platten, which is in contact only with the bottom surface of

FIGURE 3.9: *Typical commercial horizontal slab gel electrophoresis apparatus. (Courtesy of Pharmacia Ltd.)*

the gel, while the upper surface of the gel is exposed to the air. There will inevitably be a temperature gradient between the upper and lower surfaces of the gel which will be dependent on the thickness of the gel. Therefore, it is important to limit gel thickness to a maximum of 0.5 mm. Providing that an effective cooling system is employed, relatively high field strengths can be applied, leading to decreased separation times and improved protein resolution. This approach has been taken to an extreme by some groups, particularly for isoelectric focusing applications (see Section 7.2.5) where it is advantageous to use very high field strengths, and ultra-thin gels of thickness 50–100 µm [6,7].

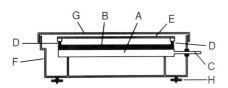

FIGURE 3.10: Diagram of typical horizontal slab gel electrophoresis apparatus. A, Cooling platten; B, polyacrylamide slab gel; C, connections to circulating chiller; D, electrodes; E, electrode support; F, electrophoresis chamber; G, lid; H, levelling feet.

Horizontal slab gels are prepared using a casting cassette in essentially the same manner as described previously for vertical slab gels. However, handling of the gels is difficult if they are cast directly onto the surface of a glass plate, unless they are irreversibly bonded with an agent such as methacryloxypropyl-trimethoxy-silane (Bind-Silane^{TM}, Pharmacia). It is much better to prepare the gels on specially treated plastic sheets such as GelBond PAG^{TM} film (Pharmacia). The cassette for gel casting is made from two glass plates, one of which is covered with GelBond PAG film by rolling it on to the plate with a few drops of water (*Figure 3.11*). The other plate bears a U-frame of the required thickness. The U-frame can be cut from sheets of Parafilm (one layer corresponds to 0.12 mm), or glass plates with permanently attached U-frames are commercially available (see Appendix B). The plate with the U-frame should be treated with Repel-Silane^{TM} to prevent the gel from sticking to the glass plate and allowing easier disassembly of the cassette after polymerization.

FIGURE 3.11: Application of GelBond PAG film to the supporting glass plate using a rubber roller.

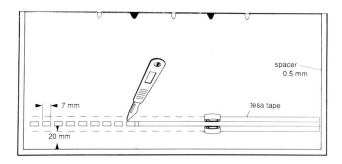

FIGURE 3.12: *Preparation of slot formers. A strip of tape applied to a glass plate is cut away using a scalpel to leave rectangles of tape around which the gel can polymerize to form sample application slots.*

At this stage slot formers can be attached to this glass plate to produce the slots within the gel which will accommodate the samples. A strip of tape applied to the plate can be cut away using a scalpel to leave rectangles of tape around which the gel will polymerize to form the slots (*Figure 3.12*). A better technique is to use silicone rubber applicator strips, which can be applied directly to the surface of the gel when it is lying on the flat-bed apparatus. The cassette assembly is then clamped together (*Figure 3.13*), placed in a vertical position and filled with the appropriate volume of acrylamide solution using a pipet (*Figure 3.14*). After polymerization is complete, the cassette should be gently prized open using a small spatula and the gel, on its plastic support, removed. Alternatively, a range of ready-made gels suitable for a variety of electrophoretic techniques is available for use with the Pharmacia horizontal electrophoresis systems (see Appendix B).

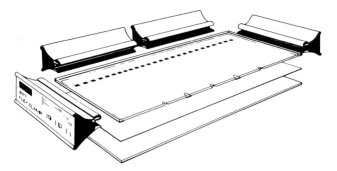

FIGURE 3.13: *Assembling a cassette for casting a horizontal slab gel.*

The popularity of horizontal slab gel systems appears to be increasing and this approach is certainly the method of choice for most applications using isoelectric focusing (see Chapter 7), agarose

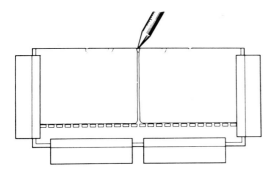

FIGURE 3.14: Casting a horizontal polyacrylamide slab gel.

electrophoresis, and immunoelectrophoresis [8]. In addition, a fully automated integrated computer-controlled electrophoresis system, known as the PhastSystem (*Figure 3.15*), has been developed [9,10] by Pharmacia. The equipment is very versatile, as all of the major gel electrophoretic techniques are supported, and it incorporates a module for automated staining of the resulting separations (*Figure 3.15*). Additional modules are available for blotting applications, and densitometry and image analysis. The gel format is very small (4.3 × 5.0 cm) so that rapid separations can be achieved. The high rate of sample throughput and the level of automation of the equipment has made it very popular in laboratories requiring rapid screening of large numbers of samples. Some examples of the various types of separations that can be performed with this equipment are shown in *Figure 3.16*.

3.8 Gel concentration

3.8.1 Homogeneous gels

As we have seen from the discussion of the pore size of polyacrylamide gel (Section 3.5), proteins are subjected to molecular sieving during their passage through the gel matrix. Thus, molecular sieving, as well as charge density, exerts a significant influence on the separation of proteins by polyacrylamide gel electrophoresis (PAGE). Careful consideration must be given to the choice of the appropriate acrylamide concentration for the optimal separation of the particular proteins to be analyzed. If too high a gel concentration is chosen the proteins may be totally excluded from the gel, or if too low a gel concentration is used the proteins will not separate and will run together with the buffer front. Theoretically, there is a gel concentration which is optimal for the resolution of a given pair of proteins (see Section 4.3.1). However, there is no single gel concentration which will give maximum separation of the components of a more complex

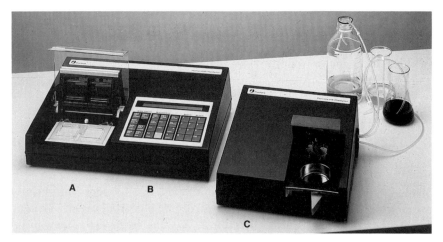

FIGURE 3.15: *Diagram of PhastSystem automated electrophoresis apparatus. A, Separation compartment; B, separation and control unit; C, staining unit. (Courtesy of Pharmacia Ltd.)*

protein mixture. A guide to gel concentrations appropriate for samples containing proteins of various size ranges is shown in *Table 3.1*. If there is no prior information on the size range expected in a particular protein sample, then a method of trial and error must be adopted, in which gels of different concentrations are tested in order to select the best concentration for the proposed study.

TABLE 3.1: *Separation ranges of acrylamide gels of various concentrations*

Gel concentration (%T)	M_r range ($\times 10^{-3}$)
3–5	> 100
5–12	20–150
10–15	10–80
> 15	< 15

3.8.2 Gradient gels

The effective separation range of polyacrylamide gels can be extended using gels containing a linear or non-linear concentration gradient. The average pore size of these gels decreases with increasing gel concentration, so that there is an effective band-sharpening effect during electrophoresis. Thus, gradient polyacrylamide gels have the dual advantages of separating proteins with a wider range of sizes and producing sharper protein zones, that is, both separation and resolution are improved. A major disadvantage of such gels, however, is the additional complexity involved in preparing gradient gels reproducibly.

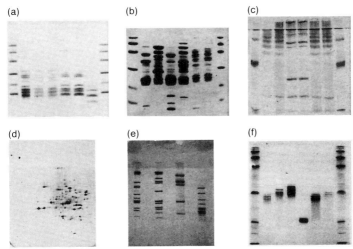

FIGURE 3.16: *Examples of various types of electrophoretic separation performed using the PhastSystem automated electrophoresis apparatus (courtesy of Pharmacia LKB Biotechnology). (a) SDS–PAGE on PhastGel gradient 10–15. Samples: E. coli ribosomal proteins and MW standards; (b) SDS–PAGE on PhastGel homogeneous 20. Samples: fractions from synthesis of human growth hormone and MW standards; (c) native PAGE with PhastGel homogeneous 7.5. Samples: Fab fragments; (d) 2-D PAGE of E. coli extract with silver staining. First dimension run on PhastGel IEF 3–9; second dimension on PhastGel gradient 10–15; (e) DNA fragments separated on PhastGel homogeneous 12.5; (f) isoelectric focusing on PhastGel IEF 3–9. Samples: murine monoclonal antibodies and pI standards.*

a. Linear gradient gels. Most separations are carried out using linear polyacrylamide gradients as their preparation is relatively straightforward. As an example of the extended separation range that can be achieved with gradient polyacrylamide gels, a 5–20%T (2.6%C) gel is able to separate proteins over the M_r range 14–200×10^3.

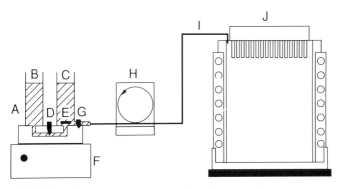

FIGURE 3.17: *Apparatus for casting linear gradient polyacrylamide gels. A, Two-chamber gradient mixer; B, reservoir (low %T); C, mixing chamber (high %T); D, valve; E, stirring bar; F, magnetic stirrer; G, stopcock; H, peristaltic pump; I, tubing; J, slab gel cassette.*

Linear polyacrylamide gels can be produced in either rod or slab formats, the latter being much more popular. The simplest method of producing such gels is to use a gel casting cassette as described in Section 3.7.2. A simple two-chamber gradient mixer is used to deliver the solution to the top of the cassette, high %T first, its flow being controlled with a peristaltic pump (*Figure 3.17*). Considerable heat is generated during gel polymerization and this results in convective disturbances which tend to break down the gradient. This effect can be minimized by stabilizing the acrylamide gradient with an accompanying density gradient of glycerol or sucrose. A more complicated solution to this problem is to generate a gradient of polymerization catalyst so that polymerization progresses from the top of the gel (low acrylamide concentration) to the bottom (high acrylamide concentration).

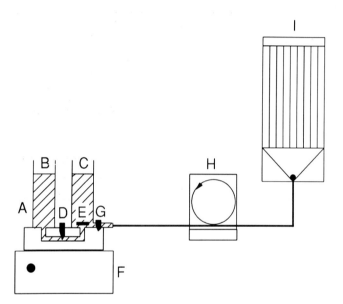

***FIGURE 3.18**: Apparatus for casting batches of linear gradient polyacrylamide gels. A, Two-chamber gradient mixer; B, reservoir (high %T); C, mixing chamber (low %T); D, valve; E, stirring bar; F, magnetic stirrer; G, stopcock; H, peristaltic pump; I, tower containing slab gel cassettes.*

A major problem is the reproducibility of gradient polyacrylamide gels. It is preferable to produce gradient gels in batches: this is achieved by placing a batch of gel cassettes in a gel-forming tower, the gel gradient solution being delivered to the bottom of the tower, low %T first (*Figure 3.18*). The acrylamide gradient can again be generated using a large volume two-chamber mixer. Greater versatility and control over gradient reproducibility can be achieved using a microcomputer to control an electromagnetic two-way valve (*Figure 3.19*). With this type of system it is necessary to interpose a small-volume mixing chamber between the valve and the peristaltic pump.

PROPERTIES OF POLYACRYLAMIDE GELS

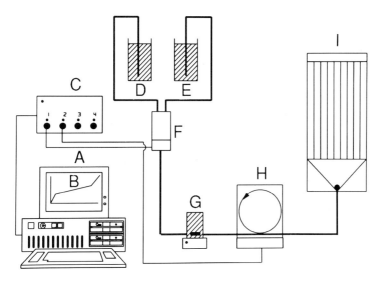

FIGURE 3.19: *Microcomputer-controlled apparatus for preparation of gradient polyacrylamide slab gels. A, Microcomputer; B, shape of gradient to be used; C, switch controller; D, reservoir for high %T acrylamide solution; E, reservoir for low %T acrylamide solution; F, electro-magnetic two-way valve; G, mixing chamber; H, peristaltic pump; I, tower containing gel cassettes.*

A range of ready-made gradient gels of different polyacrylamide concentration ranges and buffer systems are available from several commercial suppliers (see Appendix B). Such commercial gels are, of course, prepared in a highly controlled manner and should ensure reproducible separations. However, many researchers still favor the preparation of their own gels due to the extra flexibility and versatility that this provides.

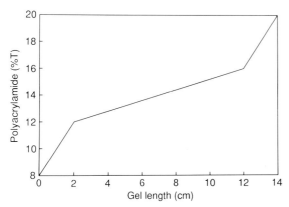

FIGURE 3.20: *A non-linear polyacrylamide gradient containing a flattened region to improve the separation of closely spaced protein zones.*

b. Non-linear gradients. For a particularly difficult separation problem it may be necessary to resort to the use of convex, concave or more complicated gradient shapes rather than a linear one. For example, a gradient containing a flattened region, as shown in *Figure 3.20*, will improve the separation of protein zones which are closely spaced in that part of the gradient. Concave gradients can be produced using simple two-chamber devices in which the cross-sectional area of the mixing chamber is either larger (low %T first) or smaller (high %T first) than that of the other chamber, or by using a gradient maker with a constant-volume mixing chamber [11]. A greater degree of flexibility can be obtained using a two-chamber system of variable complementary shape [12] or a microcomputer-controlled gradient making system.

References

1. Allen, R.C., Saravis, C.A. and Maurer, H.R. (1984) *Gel Electrophoresis and Isoelectric Focusing of Proteins*. Walter de Gruyter, Berlin.
2. Righetti, P.G. (1983) *Isoelectric Focusing: Theory, Methodology and Applications*. Elsevier, Amsterdam.
3. Rothe, G.M. and Maurer, W.D. (1986) in *Gel Electrophoresis of Proteins* (M.J. Dunn, ed.). Wright, Bristol, p. 37–140.
4. Görg, A., Postel, W., Günther, S. and Weser, J. (1986) in *Electrophoresis '86* (M.J. Dunn, ed.). VCH, Weinheim, p. 435–449.
5. Görg, A., Postel, W., Weser, J., Schiwara, H.W. and Boeksen, W.H. (1985) *Science Tools*, **32**, 5–9.
6. Görg, A., Postel, W. and Westermeier, R. (1978) *Anal. Biochem.*, **89**, 60–70.
7. Radola, B.J. (1980) *Electrophoresis*, **1**, 43–56.
8. Bog-Hansen, T.C. (1990) in *Gel Electrophoresis of Proteins: a Practical Approach* (B.D. Hames and D. Rickwood, eds). IRL Press, Oxford, p. 273–300.
9. Olsson, I., Axiö-Fredriksson U.B., Degerman, M. and Olsson, B. (1988) *Electrophoresis*, **9**, 16–22.
10. Olsson, I., Wheeler, R., Johansson, C., Ekström, B., Stafström, N., Bikhabhai, R. and Jacobson, G. (1988) *Electrophoresis*, **9**, 22–27.
11. Andrews, A.T. (1986) *Electrophoresis: Theory, Techniques and Biochemical and Clinical Applications*. Clarendon Press, Oxford.
12. Anderson, N.L. and Anderson, N.G. (1978) *Anal. Biochem.*, **85**, 341–354.

4 Electrophoresis under Native Conditions

Electrophoresis under native conditions is used in circumstances where it is desired to maintain both the subunit interactions and the native conformation of the proteins to be analyzed. This should ensure that the biological activity of the separated components is preserved, so that properties such as enzyme activity, antibody binding, and receptor activity can be studied after electrophoresis. Unfortunately, native electrophoresis techniques can only be applied to protein samples which are soluble and which will not precipitate or aggregate during electrophoresis.

4.1 Continuous buffer systems

The simplest form of PAGE involves the use of a uniform polyacrylamide concentration in conjunction with a single homogeneous buffer system. This group of techniques is known as continuous zone electrophoresis (CZE). As separation during electrophoresis will depend on both the size and charge of the proteins being analyzed, these properties should be exploited to optimize the separation of the components of interest. Selection of the appropriate gel concentration, using either homogeneous or gradient gel systems, for the size range of proteins to be separated, has already been discussed (see Section 3.8).

4.1.1 Choice of pH and buffer system

The charge properties of the proteins being analyzed can be exploited by careful choice of the pH of the buffer used for electrophoresis. Electrophoresis of proteins can be performed at any pH within the range pH 2–11, but protein deamidation and hydrolytic reactions can be significant at extreme pH values (below pH 3 and above pH 10). The stability of the proteins in the sample to be analyzed must also be considered: the use of buffer pH values at which the particular proteins are unstable or prone to precipitation must be avoided. In particular, many native proteins are liable to precipitate at pH values close to their isoelectric points (pI). Moreover, proteins bear no net charge at their pI, so the further the pH of the electrophoresis buffer from the pI values of the proteins to be separated, the greater the charge on the proteins. This results in shorter running times and reduced band spreading due to diffusion. The range of pH values at which a protein is biologically active should also be considered if functional studies are to be performed subsequent to the separation.

Many proteins have pI values in the range pH 4–7, so that the most commonly used buffer systems operate in the slightly alkaline range (pH 8–9) as the majority of proteins will be negatively charged under these conditions. In this case the samples are applied at the cathode and migrate towards the anode during electrophoresis. However, it is important to realize that in such a system any basic proteins present in the sample being analyzed will migrate in the opposite direction (i.e. towards the cathode) and be lost from the gel. If basic proteins (e.g. histones) are to be analyzed, then electrophoresis should be performed under acidic conditions where most of the protein will exist as cations. Of course, it is then necessary to apply the samples at the anodic end of the gels. Some commonly used buffer systems are given in *Table 4.1*.

TABLE 4.1: Commonly used homogeneous buffer systems for polyacrylamide gel electrophoresis

Approximate pH range	Primary buffer constituent	pH adjusted to the desired value with
2.4–6.0	0.1 M citric acid	1 M NaOH
2.8–3.8	0.05 M formic acid	1 M NaOH
4.0–5.5	0.05 M formic acid	1 M NaOH or Tris
5.2–7.0	0.05 M maleic acid	1 M NaOH or Tris
6.0–8.0	0.05 M KH_2PO_4 or NaH_2PO_4	1 M NaOH
7.0–8.5	0.05 M Na diethyl-barbiturate	1 M HCL
7.2–9.0	0.05 M Tris	1 M HCl or glycine
8.5–10.0	0.015M $Na_2B_4O_7$	1 M HCl or NaOH
9.0–10.5	0.05 M glycine	1 M NaOH
9.0–11.0	0.025 M $NaHCO_3$	1 M NaOH

Reproduced from reference [1] with permission from Oxford University Press

4.1.2 Choice of ionic strength

Ionic strength is another parameter that should be considered when choosing the buffer for a particular electrophoretic separation. As discussed in Chapter 2, buffers of low ionic strength permit higher rates of protein migration, while those of higher strength generally result in sharper zones of separation. In addition, the ionic strength must be maintained at a level where there is sufficient buffering capacity to maintain the desired pH during electrophoresis and to maintain protein solubility. However, in practice the choice of ionic strength is usually limited by considerations of Joule heating (see Section 2.4) as the more concentrated the buffer, the higher the current at a given applied voltage, and the higher the heat generated. In most systems this means that the concentration of buffer used for electrophoresis should not exceed 0.1 M.

4.1.3 Advantages and disadvantages

The main advantage of CZE procedures is that gel preparation and the electrophoretic procedure are both rapid and simple. In addition, the buffer composition and separation pH are precisely established, which can be an important consideration if the stability of the protein being analyzed is sensitive to buffer pH and/or buffer composition. The main disadvantage of these techniques is that they have an inherently rather low resolution capacity compared with other electrophoretic techniques. The method is not suitable for the analysis of dilute sample solutions since there is no concentration effect in the initial stage of the separation (see Section 4.2). Thus, large sample volumes result in broad protein zones in the final separation pattern. Acceptable and useful separations can be achieved, however, if the sample is available in a concentrated state (> 1 mg/ml) so that it can applied as a narrow starting zone.

4.2 Discontinuous buffer systems

Despite their simplicity, PAGE techniques based on CZE are seldom used nowadays and they have been largely replaced by methods using discontinuous (multiphasic) buffer systems. This group of techniques are termed multiphasic zone electrophoresis (MZE), but are generally referred to as 'disc electrophoresis'. The term 'disc' refers to the nature of the buffer system and not to the fact that this methodology was developed using cylindrical rod gels rather than the slab gels normally used today.

4.2.1 Advantages

The popularity of this group of techniques is due to their ability to concentrate the sample proteins into a narrow starting zone or stack. This results in very sharp protein zones, so that these procedures have an inherently high resolution capacity. The method is applicable to quite dilute sample solutions, but one should be aware that during the stacking phase, the concentration of the sample proteins can be so great as to induce concentration-dependent artifacts, such as irreversible protein interactions. Care should also be taken to ensure that the different pH conditions operative during stacking and separation are not such as to affect adversely the chemical, physical or biological properties of the sample proteins.

4.2.2 Mechanism of stacking

The mechanism of stacking will be discussed, using as an example the original discontinuous buffer system for PAGE which was developed by Ornstein [2] and Davis [3]. The basic mechanism of stacking is illustrated in *Figure 4.1*. A discontinuous buffer system has four important features: (i) there is a large pore (i.e. low polyacrylamide concentration) stacking gel between the sample zone and the separating gel; (ii) the separating gel has a higher acrylamide concentration appropriate to the size range of the proteins to be separated; (iii) the sample and both gels contain chloride ions, while the electrode reservoir buffer contains glycinate ions; and (iv) the pH of the resolving gel (pH 8.9) is higher than that of the sample and the stacking gel (pH 6.7).

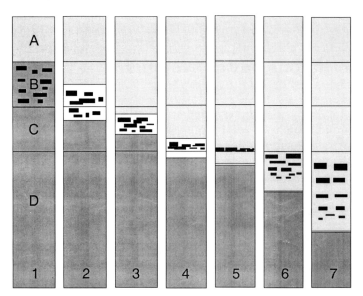

FIGURE 4.1: Mechanism of stacking. **(A)** Reservoir buffer containing glycine; **(B)** sample containing Cl^- at pH 6.7; **(C)** stacking gel containing Cl^- at pH 6.7; **(D)** separating gel containing Cl^- at pH 8.9. ☐ glycinate; ▨ Cl^-; ☐ low conductivity zone; ■ proteins.

Sample loaded on to stacking gel (1); proteins stacking in low conductivity zone (2–4); proteins unstacking in separating gel (5–7); glycinate trailing proteins (2–4), glycinate moves ahead of proteins (5–7). For further details see text.

At pH 6.7, glycinate is poorly dissociated, so that its electrophoretic mobility is low, while chloride is highly dissociated with a consequently high mobility. Sample proteins have mobilities intermediate between those of chloride and glycinate. When an electric field is applied, the rapidly migrating chloride (leading) ions move away from the slowly migrating glycinate (trailing) ions, leaving behind a zone of lower conductivity. Since conductivity is inversely proportional to field strength, a steeper voltage gradient is established in this region, thereby accelerating the trailing ions so that they migrate immediately behind the leading ions. In this way a moving boundary is formed which sweeps rapidly through the sample zone. The proteins with intermediate mobilities between the leading and trailing ions are concentrated into the boundary to form thin layers or stacks in order of their individual mobilities. It is important that the effects of sieving of sample proteins is minimized in the stacking gel. Therefore, a low concentration of acrylamide, usually about 3%T, is used to form the stacking gel.

When the moving boundary reaches the interface between the stacking and separating gels, it encounters a region of increased pH and decreased pore size. At pH 8.9, the degree of dissociation of glycinate is markedly increased, so that its mobility is nearly equivalent to that of the chloride ions and is greater than that of the sample proteins. The trailing ions, therefore, overtake the proteins to form a moving boundary between themselves and the leading ions which accelerates away from the stacked layers of sample proteins. This effect, together with the decrease in the effective pore size of the separating gel, causes the proteins to become unstacked. The proteins are then subsequently separated according to size and charge as discussed previously for continuous buffer systems (Section 4.1).

4.2.3 Choice of discontinuous buffer system

The majority of separations are carried out using the original Ornstein–Davis buffer system [2, 3]. However, it should be pointed out that this was developed specifically for the separation of serum proteins. As in the case of continuous buffer systems, there is no universal buffer system suitable for all types of sample. Thus, if the Ornstein–Davis system produces unsatisfactory results or if a particular separation is to be optimized, it is necessary to select the appropriate buffer system. Computational treatments have been developed to establish the constituents of moving boundary systems with the desired properties. The most powerful of these is that of Jovin [4] which generated in excess of 4000 buffer systems, forming the so-called 'extensive buffer system output'. A selection of 19 of the most useful of these buffer systems has been published [5] to aid the investigator in the selection of the most appropriate system for a particular separation problem. This topic is discussed in greater detail in reference [6], to which the interested reader is referred.

4.3 Estimation of molecular mass of native proteins

Native proteins are separated by electrophoresis on the basis of both their charge and size properties. Thus, if the effects of protein charge on the separation can be eliminated or taken into account, then gel electrophoresis can be used to estimate the molecular weight of the proteins.

4.3.1 Ferguson plot

There is a linear relationship between gel concentration (%T) and the logarithm of relative mobility (R_f) [7], which can be represented by:

$$\log R_f = \log Y_0 - K_R T.$$

Relative mobility is defined as the mobility of the protein of interest measured with reference to the buffer front as detected by a tracking dye (i.e. a negatively charged low molecular weight dye that migrates with the buffer front). Thus, R_f = distance migrated by protein/distance migrated by tracking dye. The intercept, Y_0, is the mobility at 0%T, that is, it gives an estimate of the mobility of the protein in free solution. The slope of the line is the retardation coefficient, K_R. Examples of different types of Ferguson plots which can be obtained are shown in *Figure 4.2*.

It has been established that there is a linear relationship between K_R and the molecular mass of native proteins [8]. A series of standard native proteins, whose molecular masses have been previously established by other procedures, can be used to construct a plot of K_R versus molecular mass. It is recommended that gels with at least seven different %T values should be used to give accurate values of K_R and Y_0 [6]. The K_R of the sample protein(s) can then be measured and the corresponding molecular mass estimated from the standard curve. A suite of computer programs, known as PAGE–PACK, has been developed to automate this procedure and are discussed in detail in reference [6].

It should be noted that K_R can depend on many factors so that when performing Ferguson analysis it is essential to keep experimental conditions (e.g. %C, pH, buffer composition, ionic strength, temperature) as constant as possible. Another problem is that measurements of mass are dependent on the shape, degree of hydration, and partial specific volume of the proteins being analyzed and those used to construct the standard curve. The presence of carbohydrate, lipid or other prosthetic groups on proteins can also cause problems. Nevertheless, this method has been widely applied and plots of K_R against molecular mass are usually linear in the range M_r 45–500 × 10^3. The methodology of Ferguson plots is discussed in more detail in reference [6].

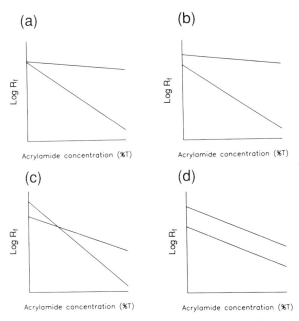

FIGURE 4.2: Examples of Ferguson plots. **(a)** The two components differ in size but not in net molecular charge; **(b)** the smaller molecule has the higher net molecular charge and free-solution mobility; **(c)** the larger molecule has the higher net molecular charge; and **(d)** the two components differ in charge but are of the same molecular size.

4.3.2 Pore gradient electrophoresis

We have seen in Chapter 3 that the effective separation range of polyacrylamide gels can be extended using gels containing linear or non-linear concentration gradients. In addition, the protein zones are subjected to a sharpening effect due to the gradient. Protein migration in a gradient gel depends on various factors (e.g. molecular mass, charge, time, gel gradient, field strength) so that there is no physical constant equivalent to relative mobility (R_f) in these systems. Gradient electrophoresis is sometimes known as 'pore limit electrophoresis' as migration rates decrease until each protein reaches its pore limit, after which further migration occurs at a slow rate proportional to time. The resulting protein profiles are consequently stable and highly resolved.

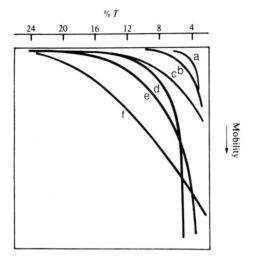

FIGURE 4.3: *Transverse gradient electrophoresis of a mixture of proteins of different sizes (**a** is the largest, **f** is the smallest) and of varying net molecular charges. An overall examination of the separation profile shows that all the consituents of this mixture could be separated from each other if a homogeneous gel of about 6.5 %T was used. Note that this value of %T is not ideal for the separation of any particular pair of proteins. Reproduced from reference [1] with permission from Oxford University Press.*

Once a protein has reached its pore limit, there is a relationship between the distance travelled by the protein and its molecular mass. Various different mathematical procedures have been developed to estimate the molecular mass of proteins using gradient gels and these are discussed in detail in reference [9].

4.3.3 Transverse gradient gel electrophoresis

In this technique a gradient gel is prepared in the normal manner. It is subsequently rotated through 90°, so that the protein sample is applied across the gradient. Thus, during electrophoresis the sample proteins migrate simultaneously into areas of different gel concentrations and form smooth continuous arcs (*Figure 4.3*). Using this approach, Ferguson plots can be constructed using a single transverse gradient gel, rather than using a series of homogeneous gels of different concentrations [10]. This method is considered to be less accurate than the traditional approach to Ferguson plot analysis; nevertheless it can be used as a rapid method for establishing the gel concentration required for the optimal separation of a particular protein in mixture.

References

1. Andrews, A.T. (1986) *Electrophoresis: Theory, Techniques and Biochemical and Clinical Applications.* Clarendon Press, Oxford.
2. Ornstein, L. (1964) *Ann. N.Y. Acad. Sci.*, **121**, 321–349.
3. Davis, B.J. (1964) *Ann. N.Y. Acad. Sci.*, **121**, 404–427.
4. Jovin, T.M. (1973) *Ann. N.Y. Acad. Sci.*, **209**, 477–496.
5. Chrambach, A. and Jovin, T.M. (1983) *Electrophoresis*, **4**, 190–204.
6. Chrambach, A. (1985) *The Practice of Quantitative Gel Electrophoresis.* VCH, Weinheim.
7. Ferguson, K. (1964) *Metabolism*, **13**, 985–1002.
8. Hedrick, J.L. and Smith, A.J. (1968) *Arch. Biochem. Biophys.*, **126**, 155–64.
9. Rothe, G.M. and Maurer, W.D. (1986) in *Gel Electrophoresis of Proteins* (M.J. Dunn, ed.). Wright, Bristol, p. 37–140.
10. Margolis, J. and Kenrick, K.G. (1968) *Anal. Biochem.*, **25**, 347–362.

5 Electrophoresis in the Presence of Additives

Gel electrophoresis of proteins under native conditions generally works well for samples of soluble proteins. However, problems arise if any of the proteins present in the sample are insoluble or are likely to form multimolecular complexes or aggregates under the conditions used for the separation. In these circumstances, effective electrophoretic separation can only be achieved if suitable additives are present in the gel to increase protein solubility and minimize aggregate formation.

5.1 Disulfide bond cleaving agents

Disulfide bonds formed between polypeptides are often involved in the formation of protein aggregates. In addition, they often play a major role in the stabilization of proteins with a subunit structure. Disulfide bonds can be readily reduced by adding reagents such as 100 mM 2-mercaptoethanol or 20 mM dithiothreitol to the sample prior to electrophoresis. Sample oxidation, with the reformation of disulfide bonds, can occur during electrophoresis. Unfortunately, disulfide bond cleaving reagents cannot be added directly to the gel mixture as they inhibit polymerization. If it is important to maintain reducing conditions within the gel during electrophoresis, low concentrations (around 1 mM) of dithiothreitol or thioglycollate can be added to the upper electrolyte reservoir.

5.2 Urea

Urea is a commonly used gel additive designed to increase sample solubility and minimize protein aggregation. Proteins unfold and are denatured progressively as they are exposed to increasing concentrations of urea, and different proteins have different sensitivities to denaturation by urea. Thus, low concentrations of urea are used if it is desired to maintain proteins in a native state, whereas high concentrations of urea (typically 8 M or higher) are used if the proteins are to be separated in a denatured state. Urea should be included in both the separating and stacking gels if a discontinuous buffer system is being used. It is not necessary to add urea to the electrolyte buffer solutions. Urea should also be added to the sample prior to electrophoresis. Indeed, it is often possible to use lower concentrations of urea in the gel phase (e.g. 4 M) provided that 8 M urea has been added to the sample.

It is important to remember that urea breaks down, particularly at alkaline pH, to form cyanate ions which can interact with the amino groups of proteins. This process of carbamylation results in charge modification of the proteins which can result in anomalous electrophoretic properties and lead to artifactual separation profiles. Cyanate formation from urea is temperature dependent, so that it is essential that protein samples containing urea are not heated. Cyanate ions can be readily removed from solutions containing urea (e.g. an acrylamide gel mix) by treatment with a suitable cationic or mixed-bed ion exchange resin as described in Section 3.3. This procedure also serves to remove charged breakdown products formed from the acrylamide (see Section 3.3).

A particularly important application of gels containing urea is for the separation of basic proteins such as histones. A useful method for the routine analysis of histones is an acid–urea gel system containing 2.5 M urea and 0.9 M acetic acid (pH 2.7) [1]. The five major histones can be separated in this system (*Figure 5.1*). This system can be improved to give increased separation and resolution of histone proteins by the addition of non-ionic detergent (see Section 5.4.1).

5.3 Transverse gradients of urea

As discussed in Section 5.2, different proteins have different sensitivities to unfolding and denaturation by urea. These conformational properties of proteins can be investigated by incorporating a transverse gradient of urea within the gel, perpendicular to the direction of migration during electrophoresis, so that a protein applied as a continuous zone across the top of the gel will migrate simultaneously into regions containing different concentrations of urea. Unfolding of the protein is detected as a decrease in electrophoretic mobility because the unfolded protein has a larger hydrodynamic volume (*Figure 5.2*). This method can be applied to complex mixtures as the various proteins are separated during electrophoresis and can be analyzed individually. This technique was developed by Creighton [2] and is discussed in detail in reference [3].

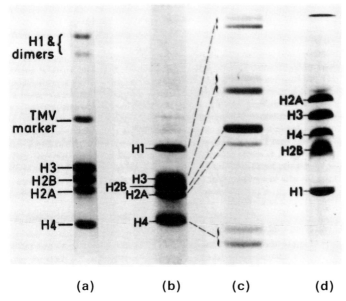

FIGURE 5.1: *Separation of histones by PAGE techniques: (a) SDS–PAGE; (b) 2.5 M urea and 0.9 M acetic acid using a 9 cm gel; (c) 2.5 M urea and 1 M acetic acid using a 25 cm gel; and (d) 2.5 M urea, 0.9 M acetic acid and 0.4% Triton X-100 using a 9 cm gel. Reproduced from reference [17] with permission from IRL Press.*

5.4 Detergents

A popular and effective approach to the solubilization and disaggregation of proteins to be analyzed by electrophoresis is to include a detergent both in the sample and in the gel. Detergents can be classified into four general categories: (i) non-ionic, (ii) amphoteric or zwitterionic, (iii) anionic, and (iv) cationic. Examples of some commonly used detergents in each of these categories are shown in *Figure 5.3*. It should be appreciated that detergents are most effective at disrupting hydrophobic interactions involved in protein–lipid and protein–protein interactions. They are generally ineffective in disrupting ionic or polar interactions and, obviously, have no effect on disulfide bond formation.

A major factor that must be considered when choosing a detergent appropriate for a particular electrophoretic separation is whether the biological properties and/or enzymatic activities of the separated

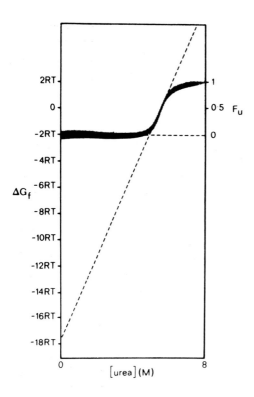

FIGURE 5.2: *Urea gradient gel electrophoresis of horse ferricytochrome c. The migration of the protein in the gel (solid black line) illustrates the decrease in electrophoretic mobility which occurs as the protein unfolds with increasing urea concentration. F_u is the fraction of the protein which is unfolded (right scale). Near the midpoint of the transition (folded to unfolded), where $F_u = 0.5$, the net stability (ΔG_f) (left scale) is zero. The unfolded and native states correspond to $\Delta G_f = 2RT$ and $-2RT$, respectively. The extrapolation of ΔG_f from the transition region to zero urea concentration gives the net stability in the absence of urea. Reproduced from reference [3] with permission from Academic Press. For more information see reference [3].*

proteins are to be preserved. The anionic and cationic detergents are generally considered to be strongly denaturing, while many non-ionic detergents are regarded as being mild and non-denaturing. However, a distinction must be made here between the strict physical chemistry definition of denaturation, which denotes the disruption of the tertiary structure of a protein to produce a random coil configuration, and the looser definition used in biochemistry, where it includes deleterious effects on binding properties and enzymatic activity. Certain detergents which would be considered as mild in their effects on molecular structure are nevertheless able to exert effects on the biological activities of proteins. Thus, a process of trial and error must normally be used to determine the best detergent for a particular electrophoretic separation. This topic and other important aspects of the use of detergents in electrophoresis are discussed in reference [4]. Also, the field of membrane protein biochemistry has given a major impetus to the search for detergents with effective solubilization properties and which are compatible with gel electrophoresis (reviewed in references [5] and [6]).

5.4.1 Non-ionic detergents

Most of these detergents are considered to be mild and are often used when the preservation of biological or enzymatic activity of the separated proteins is required. Another important feature of non-ionic detergents is that they do not result in modification of the net charge of the protein so that electrophoretic separation will still be dependent on the native charge and size of the proteins being analyzed. This also means that these detergents are compatible with the important group of techniques using isoelectric focusing (see Section 7.2.3). Non-ionic detergents are normally included in samples and gels in the range 0.1–5.0% (w/v). They are also compatible with gels containing high concentrations of urea.

The structures of some commonly used non-ionic detergents are shown in *Figure 5.3*. There is a certain amount of confusion in this field: although there are relatively few general classes of detergent, they are available commercially under a variety of trade names [4]. For example, Triton X-100 and Nonidet P-40, both commonly used in gel electrophoresis, are essentially equivalent detergents. Another factor to bear in mind is that most of these detergents are produced for use in domestic products such as shampoo and washing-up liquid. Therefore, the purity of some of these reagents can be questionable and it is recommended that detergents for electrophoretic applications are purchased from suppliers who have quality-tested the products for use in these techniques (see Appendix B).

The most popular and widely used non-ionic detergents are those of the polyoxyethylene ether type (e.g. Triton X-100, Lubrol PX and WX, Brij 36-T). However, other types of non-ionic detergent are available and, in particular, octyl glucoside has been found to be an effective solubilizing agent for membrane proteins [7].

Non-ionic detergents such as Triton X-100, Triton DF-16 or Lubrol WX can also be added to acid–urea gel systems for the separation of very basic proteins such as histones [8, 9]. When proteins are electrophoresed as cations at low pH in the presence of non-ionic detergents their separation depends on their ability both to bind the detergent and to form mixed micelles between the detergent and the hydrophobic domains of the polypeptide chains [8]. Thus, mobility depends on protein hydrophobicity in addition to size and charge.

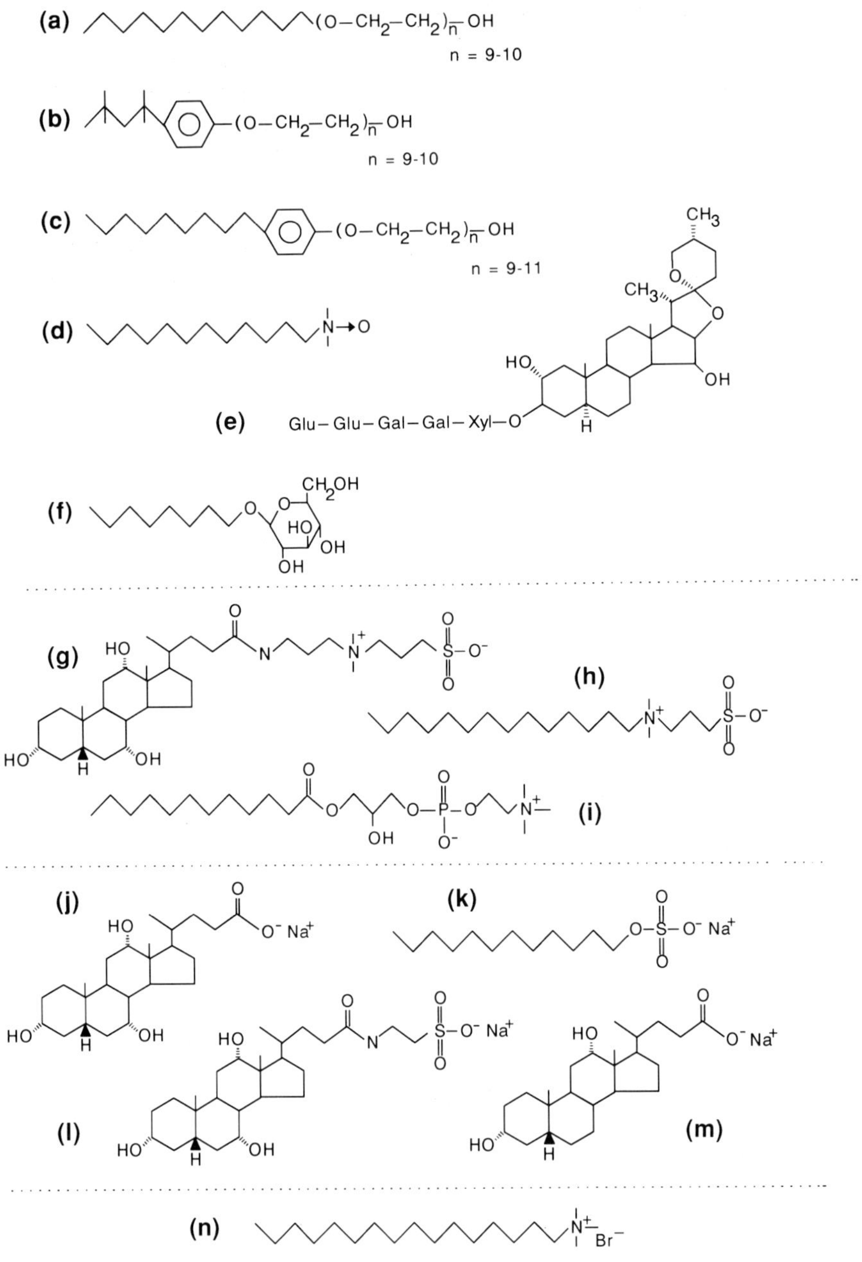

FIGURE 5.3: Structures of some common detergents. **(a)–(f)**, non-ionic; **(g)–(i)**, zwitterionic; **(j)–(m)**, anionic; **(n)**, cationic. **(a)** Lubrol PX; **(b)** Triton X-100; **(c)** Triton N-101; **(d)** Ammonyx LO; **(e)** digitonin; **(f)** octyl glucoside; **(g)** CHAPS; **(h)** Zwittergent 3-14; **(i)** lysophosphatidylcholine; **(j)** sodium cholate; **(k)** sodium dodecyl sulfate (SDS); **(l)** sodium taurocholate; **(m)** sodium deoxycholate (DOC); **(n)** cetyltrimethylammonium bromide. Adapted from reference [4] with permission from VCH.

5.4.2 Zwitterionic detergents

Zwitterionic or amphoteric detergents (*Figure 5.3*) are also used in certain electrophoretic applications. These detergents are generally considered to be milder than anionic or cationic detergents and, like non-ionic detergents, have no effect on net protein charge.

CHAPS (3-[(cholamidopropyl)dimethylammonio]-1-propanesulfonate) has been found to be particularly effective for the solubilization and analysis of membrane proteins [10, 11]. Its compatibility with gels

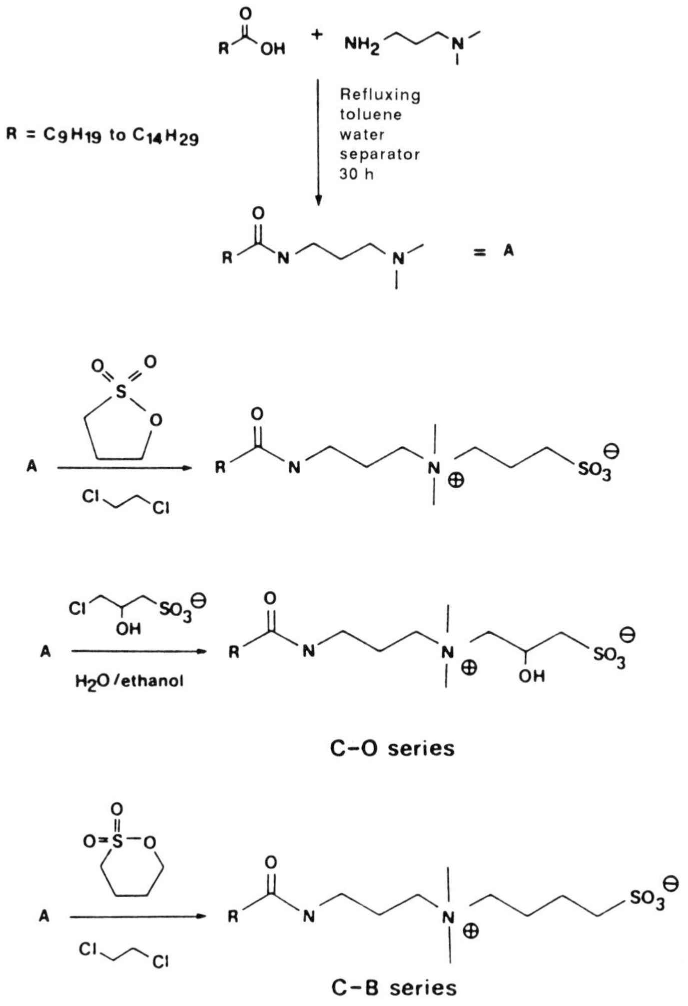

FIGURE 5.4: *Schematic diagram of the synthesis of amidosulfobetaines. The starting fatty acid is condensed with a diamine by azeotropic removal of water with toluene to give amidoamine A. This is then reacted with various reagents to give sulfobetaines. Reproduced from reference [14] with permission from Academic Press.*

containing high concentrations of urea has resulted in CHAPS being a popular, and sometimes more effective, alternative to non-ionic detergents in the isoelectric focusing dimension of two-dimensional PAGE procedures [12] (see Chapter 8).

Zwitterionic detergents such as the alkylbetaine or sulfobetaine (e.g. Zwittergent, *Figure 5.3*) type are very effective solubilizing agents. They are generally mild in their action, resulting in good retention of biological and enzymatic activity [13]. However, linear sulfobetaines are not compatible with gels containing high levels of urea as they are precipitated in the presence of concentrations greater than 4 M. To overcome this problem, a series of detergents have been developed [14] which retain the main features of classical sulfobetaines (i.e. a linear hydrophobic tail and a zwitterionic head group), but which contain a hydrophilic amido spacer in order to improve their solubility properties and thus their urea compatibility. The synthesis of these detergents is outlined in *Figure 5.4* and results in three series of compounds (B, C, and O series). It was found [14] that the best compromise between urea compatibility and solubilization performance was achieved with an alkyl chain length of 11–12 carbon atoms, with the short polar heads (C and O series) performing better than the longer head groups (B series).

5.4.3 Anionic detergents

The most important of this group of reagents is undoubtedly sodium dodecyl sulfate (SDS), sometimes called sodium lauryl sulfate (*Figure 5.3*). This is due to the ability of SDS to solubilize, dissociate, and denature the majority of oligomeric proteins into their constituent subunits. In addition, the resulting SDS–polypeptide complexes migrate towards the anode during electrophoresis in gels containing SDS in accordance with the molecular size of the individual polypeptide chains. These properties have made SDS–PAGE the most commonly used procedure for the analysis of proteins (see Chapter 6). Nevertheless, several other anionic detergents, such as sodium deoxycholate (DOC), have been used successfully for the analysis of proteins by gel electrophoresis (*Figure 5.3*).

5.4.4 Cationic detergents

The cationic detergent, cetyltrimethylammonium bromide (CTAB), has been used to study highly acidic (e.g. ferredoxins) or basic (e.g. protamines) proteins which cannot be separated satisfactorily by SDS–PAGE [15]. CTAB is simply used as a replacement for SDS, with migration being towards the cathode rather than towards the anode with SDS. Unfortunately, the resulting gels are opaque due to precipitation of cetyltrimethylammonium persulfate within the gel matrix. This problem can be overcome either by pre-electrophoresis to remove residual persulfate from the gels or by the use of alternative polymerization catalysts (e.g. photopolymerization with riboflavin) [16].

References

1. Panyim, S. and Chalkley, R. (1969) *Arch. Biochem. Biophys.*, **130**, 337–346.
2. Creighton, T.E. (1979) *J. Molec. Biol.*, **129**, 235–264.
3. Goldenberg, D.P. (1989) in *Protein Structure: a Practical Approach* (T.E. Creighton, ed.). IRL Press, Oxford, p. 225–250.
4. Hjelmeland, L.M. and Chrambach, A. (1981) *Electrophoresis*, **2**, 1–11.
5. Helenius, A. and Simons, K. (1975) *Biochim. Biophys. Acta*, **415**, 29–79.
6. Tanford, C. and Reynolds, J.A. (1976) *Biochim. Biophys. Acta*, **457**, 133–170.
7. Baron, C. and Thompson, T.E. (1975) *Biochim. Biophys. Acta*, **382**, 276–285.
8. Franklin, S.G. and Zweidler, A. (1977) *Nature*, **266**, 273–275.
9. Bonner, W.M., West, M.P. and Stedman, J.D. (1980) *Eur. J. Biochem.*, **109**, 17–23.
10. Hjelmeland, L.M., Nebert, D.W. and Osborne, J.C. (1983) *Anal. Biochem.*, **130**, 72–82.
11. Perdew, G.H., Schaup, H.W. and Selivonchick, D.P. (1983) *Anal. Biochem.*, **135**, 453–455.
12. Dunn, M.J. (1987) in *Advances in Electrophoresis* (A. Chrambach, M.J. Dunn and B.J. Radola, eds). VCH, Weinheim, **Vol. 1**, p. 1–109.
13. Gonenne, A. and Ernst, R. (1978) *Anal. Biochem.*, **87**, 28–38.
14. Rabilloud, T., Gianazza, E., Cattò, N. and Righetti, P.G. (1990) *Anal. Biochem.*, **185**, 94–102.
15. Williams, J.G. and Gratzer, W.B. (1971) *J. Chromatogr.*, **57**, 121–125.
16. Eley, M.H., Burns, P.C., Kannapell, C.C. and Campbell, P.S. (1979) *Anal. Biochem.*, **92**, 411–419.
17. Hames, B.D. (1990) in *Gel Electrophoresis of Proteins: a Practical Approach* (B.D. Hames and D. Rickwood, eds). IRL Press, Oxford, p. 1–147.

6 Sodium Dodecyl Sulfate–Polyacrylamide Gel Electrophoresis (SDS–PAGE)

SDS–PAGE is currently the most commonly used electrophoretic technique for the analysis of proteins. This is due to the ability of the strong anionic detergent SDS, when used in the presence of disulfide bond cleaving reagents (see Section 5.1), to solubilize, denature, and dissociate most proteins to produce single polypeptide chains. The resulting SDS–protein complexes can then be separated according to molecular size by electrophoresis in gels containing SDS.

6.1 Basic principles

The majority of proteins can bind 1.4 g SDS/g protein [1], effectively masking the intrinsic charge of the polypeptide chains, such that net charge per unit mass becomes approximately constant. Subsequent electrophoretic separation is dependent only on the effective molecular radius [2], which approximates to relative molecular mass (M_r), and occurs solely as a result of molecular sieving through the gel. Although there are a number of exceptions (see Section 6.6), this relationship between M_r and migration during SDS–PAGE holds true for a very large number of proteins [3].

It should be noted that many bulk commercial preparations of SDS may contain impurities, particularly materials with other chain lengths (e.g. 14 and 16 carbon chains), which can interfere with electrophoresis. It is, therefore, important to obtain a high grade reagent, preferably one which has been quality tested for electrophoresis.

6.2 Choice of gel concentration

The polyacrylamide gel concentration used determines the effective separation range of SDS–PAGE. For example, a 5%T gel will separate proteins in the M_r range 20–350 × 10^3, while a 10%T gel is suitable for proteins in the range 15–200 × 10^3. If the sample to be analyzed by SDS–PAGE contains proteins with a wide range of molecular masses, it is then advantageous to use a gradient gel system (see Section 3.8.2). For example, a 3–30%T gradient gel can separate proteins in the M_r range 10 000 to nearly 500 000.

Small polypeptides and peptides (< 10 000) cannot be separated by the standard SDS–PAGE procedure. This failure results from the fact that small molecules form SDS–protein complexes of the same dimension and charge, which then migrate together during SDS–PAGE. This problem can be overcome by adding a solute such as 8 M urea which decreases the size of the detergent micelles [4,5]. These modified SDS–PAGE systems are capable of separating polypeptides and peptides in the M_r range 1–45 × 10^3.

6.3 Buffer systems

6.3.1 Continuous buffer systems

Sodium phosphate, pH 7.2, was used as the buffer in the original SDS–PAGE procedure developed by Weber [3]. This method is effective, rapid, and simple, but suffers from the usual drawbacks of continuous buffer systems (discussed in Section 4.1.3). In particular, no concentration of the sample occurs during electrophoresis so that, unless a concentrated protein sample is applied, broad protein zones are formed with consequent loss of resolution.

6.3.2 Discontinuous buffer systems

Discontinuous buffer systems are almost universally used nowadays for SDS–PAGE, since this approach results in sharper protein zones with consequently enhanced resolution. The mechanism of stacking in multiphasic buffer systems has been described in Section 4.2.2. However, it is important to note that the nature of stacking is somewhat altered in the presence of SDS. Since SDS-coated proteins have a constant charge to mass ratio, that is, all proteins and their subunits have a uniform charge density, they will migrate with the same mobility and thus will automatically stack. Moreover, as the net charge on SDS–protein complexes does not vary between pH 7 and pH 10, mobility is not affected within this pH range. It is, therefore, not strictly necessary to have a discontinuity in pH and unstacking of the proteins will occur by the change in gel concentration [6]. Also, the observation that SDS migrates with a mobility higher than

SDS–protein complexes in a restrictive gel [6], means that SDS will overtake zones of protein in the resolving gel provided that it is included in the sample and upper buffer reservoir. Thus, SDS can be omitted from both the stacking and separating gels [7], although few investigators seem to have taken advantage of this option.

The buffer system of Laemmli. The most commonly used buffer system for SDS–PAGE is that devised by Laemmli [8], which is based on the discontinuous buffer system developed by Ornstein [9] and Davis [10] for electrophoresis of native proteins (see Section 4.2.2), but with the addition of 0.1% (w/v) SDS.

The buffer system of Neville. The discontinuous buffer system for SDS–PAGE developed by Neville [11] is based on the use of a borate–sulfate buffer system and was selected from the series of 4269 multiphasic buffers calculated by Jovin [12] (see Section 4.2.3). It is designed to stack all SDS-saturated proteins, while leaving behind unstacked all partially saturated SDS–protein complexes.

6.4 Sample preparation

It is important that samples are treated in such a way as to ensure optimal reaction with SDS. This is achieved by heating the sample made up in the appropriate sample buffer for at least 3 min at 100°C. In the Laemmli system the sample buffer contains 0.125 M Tris buffer, pH 6.8 (i.e. the same pH as the stacking gel in which the wells for sample application are formed), 2% (w/v) SDS, and 5% (v/v) 2-mercaptoethanol to cleave any disulfide bonds. The latter reagent can be replaced by 20 mM dithiothreitol. The solution also contains 10% (w/v) glycerol, to raise the density of the sample so that it can be easily applied into the sample well beneath the electrode buffer, and 0.001% (w/v) Bromophenol blue tracking dye, which migrates with and so marks the buffer front during electrophoresis.

Solid samples can be solubilized directly in the sample solution. Liquid or suspension samples should be diluted 1:1 with a double-strength preparation of the sample buffer. Care must be taken not to overload the gel system, and the total amount of sample protein that can be applied will depend on the protein concentration of the sample, its complexity, and the relative abundance of particular species. Therefore, the optimal loading for any particular sample must be determined empirically, but as a rule of thumb a loading of 1–10 µg of each polypeptide will give good results using Coomassie brilliant blue staining (see Section 9.4). For a complex protein mixture, a sample containing 50–100 µg total protein is usually sufficient. However, with silver staining (see Section 9.6) as little as 10–15 µg can be used. The total volume of sample that can be applied will, of course, be determined by the capacity of the sample wells being used.

Certain samples may be too dilute for direct analysis by SDS–PAGE, so they will require concentration prior to analysis. Techniques such as lyophilization, ammonium sulfate precipitation, dialysis against a high concentration of polyethylene glycol, and precipitation with 10% (w/v) trichloroacetic acid (TCA) can be used. The latter procedure is simple and rapid, but samples can be difficult to redissolve after TCA precipitation, which is usually due to a failure to neutralize the acidic pellet. A recommended procedure which avoids the need for neutralization is to precipitate the proteins with 5 vol. of cold acetone for 10 min at $-20°C$. The resulting protein pellet can be readily washed, dried, and redissolved in SDS sample buffer.

It is important that any insoluble residue should be removed from the samples using a micro-centrifuge prior to electrophoresis as such material will cause streaking of the protein pattern.

In some cases, such as for the analysis of immunoprecipitates and non-histone proteins, it is essential to add 8 M urea to the sample solubilization solution to ensure thorough denaturation and solubilization. In this case, there is no need to add glycerol to the solution as the urea will increase the density of the sample sufficiently.

6.5 Estimation of molecular mass by SDS–PAGE

6.5.1 Homogeneous gels

As we have seen in Section 6.1, the majority of polypeptides, when heated in the presence of SDS and a disulfide bond cleaving reagent, bind SDS in a constant mass ratio. The resulting SDS–protein complexes have identical charge densities and migrate during electrophoresis according to their M_r. It has been established that under these conditions, a plot of $\log_{10} M_r$ versus R_f (distance moved by the protein divided by distance moved by the tracker dye) shows a straight line relationship [2,3]. In order to calibrate the gel system, a set of marker polypeptides of known molecular mass are subjected to electrophoresis by SDS–PAGE on the same gel as the samples to be analyzed.

Suitable kits of marker proteins covering various molecular mass ranges are available from several suppliers. Marker proteins are usually visualized together with the sample proteins by staining after electrophoresis. Radioactive molecular mass standards are also available if autoradiographic or fluorographic detection is to be used. Prestained and colored proteins are also available, so that the migration of these markers can be observed during electrophoresis and they do not require visualization prior to measurement of R_f. However, derivatization of these proteins can result in changes to their molecular mass, so that calibration curves generated using this type of marker should be treated with caution.

After electrophoresis is complete, the marker and sample protein separations are visualized appropriately (*Figure 6.1*) and the position of the tracking dye front is marked. The R_f values for the marker proteins are estimated to construct a standard curve of $\log_{10} M_r$ versus R_f, although for convenience, it is more common to plot M_r using two- or three-cycle semi-log graph paper (*Figure 6.2*). The R_f values of the sample protein zones of interest are then used to determine their respective M_r values from the standard curve. It is important to appreciate that for a gel of any particular gel concentration (%T), the relationship between M_r and R_f is linear only over a limited range of molecular mass.

6.5.2 Gradient gels

As we have discussed in Section 3.8.2, gels containing a linear or non-linear acrylamide concentration gradient can be used to extend the range over which proteins of different molecular masses can be effectively separated. Using gradient SDS–PAGE, M_r values are determined from the following relationship:

$$\log_{10}M_r = a \log_{10}T' + b$$

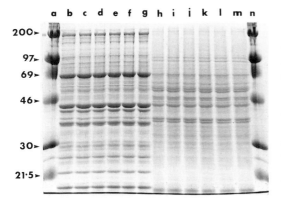

FIGURE 6.1: *10%T SDS–PAGE gel separation of total proteins of human heart (lanes b–g) and the endothelial cell line, EAHY926 (lanes h–m). Lanes a and n are M_r marker proteins. The scale at the left indicates $M_r \times 10^{-3}$. The continuous line at the bottom of the gel is the tracker dye.*

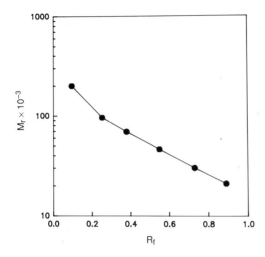

FIGURE 6.2: Calibration curve of M_r (log scale) vs R_f plotted using the M_r markers separated on the 10% T SDS–PAGE shown in Figure 6.1. Note the deviation from linearity at high M_r.

where T' is the concentration of acrylamide reached by the protein zone of interest and a and b are the slope and intercept, respectively, of the linear regression line which is established from measurements of T' for a number of standard proteins of known molecular mass run at the same time on the same gradient SDS–PAGE gel. This relationship holds true and results in linear plots of $\log_{10} M_r$ vs. $\log_{10} T'$ for gels with any shape of acrylamide concentration gradient [13,14]. In *Figure 6.3*, M_r and %T' are more conveniently plotted directly onto log–log graph paper. When the gel concentration gradient is linear, a plot of $\log_{10} M_r$ vs. $\log_{10}$ distance migrated is also linear.

The practical difficulty with the use of gradient gels for molecular mass determination is that the accuracy of the method depends on a knowledge of the precise shape of the gradient, such that T' can be

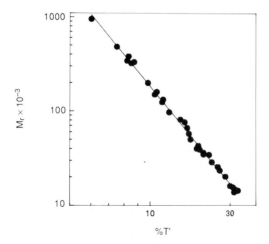

FIGURE 6.3: Calibration curve of M_r vs %T' (both log scales) for a series of 34 standard proteins covering the M_r range $13-95 \times 10^3$ separated on a 3–30%T (8.4%C) linear gradient SDS–PAGE slab gel. Plotted from data given in reference [23].

accurately estimated for each protein band. In the case of a linear gradient, T' can be accurately calculated by measurement of the distance migrated by the protein zone. This is obviously more difficult using non-linear gradients unless a highly reproducible and controllable system has been used to generate the gradient (see Section 3.8.2). The range of M_r values over which there is a linear relationship between $\log_{10}M_r$ and $\log_{10}T'$ depends on the precise acrylamide gradient employed. Small polypeptides (<13 000) can be analyzed if high concentrations of urea are added to a gradient gel system [15]. Alternatively, it has been found [16] that small polypeptides can be effectively separated using a tricine, rather than glycine, discontinuous buffer for SDS–PAGE. This procedure is able to separate polypeptides with M_r values as low as 1000.

An alternative approach for determining M_r using linear gradient SDS–PAGE gels has been developed by Rothe [17]. A straight line plot is produced from the relationship:

$$\log_{10}M_r = -a\sqrt{D} + b,$$

where D is the migration distance (mm), a the slope and b the intercept of the straight line (*Figure 6.4*). Linearity is independent of the buffer system, the concentration of cross-linker within the range 1–8%C, and the concentration range of the gradient within 3–30%T at a gel length between 8 and 15 cm. However, the values of a and b are altered by changes in these parameters. The linear relationship between $\log_{10}M_r$ and $\sqrt{D}$ is, in practice, time independent, so that estimations of molecular mass can be made when the optimal resolution for the particular protein sample being analyzed has been obtained.

6.6 Limitations of SDS–PAGE

The validity of SDS–PAGE, particularly if it is used as a method for estimation of M_r rather than for analytical separation, depends on SDS being bound to the proteins of interest in a constant mass ratio (i.e. 1.4 g SDS/g protein – see Section 6.1). While this has been shown to hold true for the great majority of proteins, there are exceptions, so that certain proteins behave anomalously in SDS–PAGE. In particular, proteins which are conjugated with other non-proteinaceous material (e.g. glycoproteins, lipoproteins) often cannot be saturated with SDS as the non-protein moieties do not react with SDS. The lower SDS binding results in a decreased charge to mass ratio. This results in a decreased mobility during electrophoresis so that the protein has a higher apparent M_r. In the case of glycoproteins this problem can be minimized by substituting the Tris–glycine buffer system of Laemmli [8] with a Tris–borate–EDTA buffer [18]. Here it is proposed that the formation of

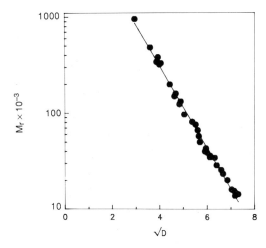

FIGURE 6.4: The method of Rothe [17] for determining M_r using linear gradient SDS–PAGE. The plot shows the linear relationship between M_r (log scale) and $\sqrt{D}$ for 34 standard proteins covering the M_r range $13–950 \times 10^3$ separated on a 3–30%T (8.4%C) linear gradient SDS–PAGE gel. Plotted from data given in reference [23].

charged borate–carbohydrate complexes results in an increase in net negative charge which offsets the decreased binding of SDS, leading to a charge density consistent with an electrophoretic migration of glycoproteins that correlates with their M_r [18]. In addition, there is some evidence that glycoproteins do not behave anomalously using gradient SDS–PAGE gels [17], so that such systems might be preferable for the analysis of this group of proteins.

Proteins with an unusual amino acid composition can also behave anomalously in SDS–PAGE. For example, collagenous polypeptides and other proteins with a high proline content exhibit anomalously high M_r values by SDS–PAGE. Proteins with a very high proportion of negatively or positively charged amino acids (e.g. histones) can also cause problems as the bound SDS does not completely mask their natural charge.

6.7 SDS–PAGE in non-reducing conditions

Samples to be analyzed by SDS–PAGE are routinely treated with agents which disrupt inter- and intramolecular disulfide bonds. For multimeric proteins stabilized by disulfide bonds, this step is essential to release the individual polypeptide subunits. The cleavage of disulfide bonds occurring within an individual polypeptide chain is also necessary to ensure complete unfolding and optimal interaction with SDS. However, SDS–PAGE of unreduced proteins can give insights into protein structure. This is best achieved by running samples of the same protein mixture, either reduced or non-reduced, applied in adjacent wells of the same SDS–PAGE slab gel (*Figure 6.5*).

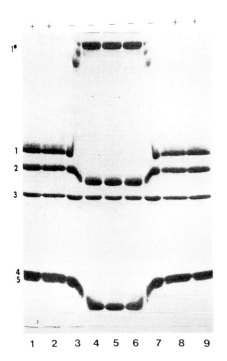

FIGURE 6.5: SDS–PAGE profiles of (1) IgG3 heavy chain, (2) IgG1 heavy chain, (3) pig skeletal muscle actin, (4) kappa light chain, and (5) kappa light chain derived from reduced IgG3k. The samples were either non-reduced (3–7) or reduced (lanes 1, 2, 8, 9) with 5% 2-mercaptoethanol prior to electrophoresis. In the absence of reducing conditions, the IgG3 heavy chain and kappa light chain migrate together as intact IgG3k (1*). Reproduced from reference [24] with permission from Academic Press.

Interpretation of results using this method can be problematical, particularly if a complex protein mixture is being analyzed. This has been overcome by the development of a diagonal SDS–PAGE method [19]. In this method (*Figure 6.6*), the non-reduced sample proteins are run in a first-dimension SDS–PAGE gel, which can be either a tube gel or a lane excised from a slab gel. After electrophoresis, the proteins are reduced *in situ* to cleave the disulfide bonds and convert cross-linked subunits into monomers. The first gel is rotated through 90° and then becomes the origin of the second slab SDS–PAGE. Polypeptides which were not involved in disulfide bridging will appear to migrate on the diagonal. Polypeptides which contained intramolecular disulfide bonds will migrate more rapidly in the second dimension and, therefore, fall below the diagonal. Multi-subunit proteins stabilized by disulfide bonds will form multiple spots off the diagonal (*Figure 6.6*).

6.8 Peptide mapping by SDS–PAGE

6.8.1 The need for peptide mapping

Electrophoretic techniques have an excellent capacity for separating and resolving complex mixtures, but do not necessarily give the investigator direct information on whether protein bands observed

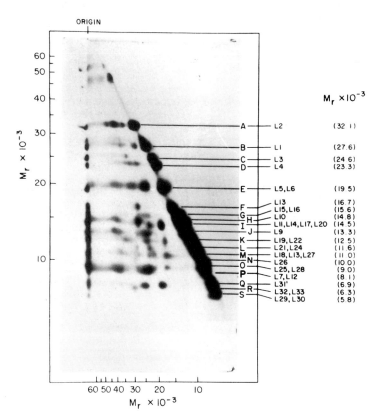

FIGURE 6.6: Diagonal gel pattern for total protein from cross-linked 50S ribosomal subunits showing the location and molecular weight of each component. 50S ribosomal proteins are numbered from L1 to L34. Reproduced from reference [19] with permission from Oxford University Press.

in different samples are related. The fact that two proteins have the same apparent mobility by electrophoresis does not mean that they are identical or even related, while polypeptides having very different mobilities may nevertheless be related.

Peptide mapping is a very powerful approach to the study of relationships between proteins. This involves cleavage of the intact protein into peptide fragments by hydrolysis of specific peptide bonds using either chemical or enzymic methods. If the proteins to be investigated are sufficiently pure, then they can be simply cleaved and the resulting peptides separated by an appropriate technique (e.g. gel electrophoresis or HPLC). Comparison of the peptide maps will indicate the degree of similarity between the proteins.

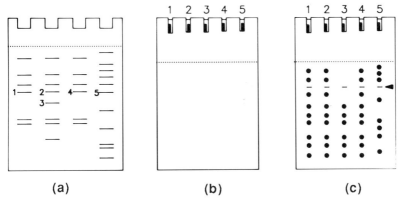

FIGURE 6.7: Representation of the key steps in an in situ peptide mapping experiment. (**a**) Protein samples are separated by gel electrophoresis and stained with Coomassie brilliant blue R-250. (**b**) Bands containing proteins of interest (1–5) are cut out of the gel, equilibrated and placed on a second gel together with a suitable protease. (**c**) The proteins and protease are then run into the gel and the resulting peptides are visualized by silver staining. The position of the protease is indicated by the arrowhead. Analysis of the peptide maps shows that proteins 1, 2, and 4 are homologous, while protein 3 is a specific degradation product of these proteins. Protein 5 has no significant homology to the other proteins. Redrawn from reference [21] with permission from IOP Publishing Ltd.

6.8.2 *In situ* peptide mapping

In many cases purified protein samples are not available, yet the investigator requires information on the similarity between individual protein bands obtained by gel electrophoresis. In order to solve this problem, a technique of *in situ* peptide mapping has been developed [20]. The steps in the procedure are outlined in *Figure 6.7* and the methodology is described in detail in references [21] and [22].

6.8.3 Primary gel system

The protein samples to be compared by peptide mapping are separated by gel electrophoresis in the normal way. Any gel electrophoretic method is compatible with this approach – native electrophoresis, SDS–PAGE, isoelectric focusing (IEF), two-dimensional (2-D) PAGE – and

TABLE 6.1: Some proteolytic enzymes used for in situ peptide mapping

Enzyme	Preferred cleavage site
Staphylococcus aureus V8 protease	Asp-X, Glu-X
Trypsin	Lys-X, Arg-X
α-Chymotrypsin	Trp-X, Tyr-X, Phe-X, Leu-X
Subtilisin	Broad specificity
Papain	Lys-X, Arg-X, Leu-X, Gly-X
Pronase	Broad specificity
Ficin	Lys-X, Arg-X, Leu-X, Gly-X

the appropriate method should be chosen to give optimal resolution of the proteins to be compared. After visualization of the protein pattern by staining or autoradiography, the protein bands to be compared are cut out from the gel. The excised bands can either be processed directly or stored frozen at $-70°C$ for up to 6 months.

6.8.4 Protein cleavage

Gel pieces are equilibrated for 30 min in a suitable buffer, normally 0.125 M Tris, pH 6.8, 0.1% (w/v) SDS, and 0.1 M EDTA, prior to fragmentation. Both chemical and enzymic methods can be used for *in situ* peptide mapping, but procedures using proteases are the most popular. Some of the enzymes which have been used together with their preferred cleavage sites are listed in *Table 6.1*. Of these, *Staphylococcus aureus* V8 protease has proved the most popular. This enzyme has a pH optimum of 8.0, but retains considerable activity at pH 6.8 and is not inactivated by SDS. Of course, V8 protease will only be effective if the proteins to be analyzed contain the appropriate cleavage sites. If they do not, then one of the other enzymes listed in *Table 6.1* must be selected. Alternatively, one of the chemical cleavage methods listed in *Table 6.2* can be used. Protein fragmentation is accomplished *in situ* in the gel pieces by incubating them under the appropriate conditions.

TABLE 6.2: Some chemical cleavage methods used for in situ peptide mapping

Reagent	Cleavage site
Cyanogen bromide	Met-X
BNPS-skatole	Trp-X
N-Chlorosuccinimide	Trp-X
Hydroxylamine	Asn-Gly
Formic acid	Asp-Pro

6.8.5 Secondary gel system

SDS–PAGE gels of the Laemmli type [8] are generally used for the peptide mapping step. These are prepared in the normal manner, except that a longer (5 cm) stacking gel is used. A 15%T gel will produce good results for the majority of peptide digests, but the gel concentration can be varied or gradient gels used to optimize the separation of any particular digest.

If enzymic cleavage is being used, the gel pieces are applied directly into the sample wells containing stacking gel buffer (0.125 M Tris–HCl, pH 6.8, 0.1% (w/v) SDS, and Bromophenol blue). The protease, made up in the same buffer containing 10% (w/v) glycerol, is then applied to

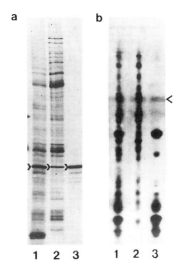

***FIGURE** 6.8:* Peptide maps of nuclear lamin c and vimentin. Three different nuclear protein fractions (a, lanes 1–3) were separated by 8%T SDS–PAGE and stained with Coomassie brilliant blue R-250. Proteins indicated by the symbol > in (a) were cut out of the gel, equilibrated and rerun together with 50 pg of V8 protease on a 15%T SDS–PAGE gel (b, lanes 1–3). The resulting peptide maps were visualized by silver staining. Analysis of the maps indicates that band (a)1 is a mixture of two proteins (lamin c and vimentin), while bands (a)2 and (a)3 represent lamin c and vimentin, respectively. The position of the V8 protease in (b) is indicated by <. Reproduced from reference [21] with permission from IOP Publishing Ltd.

the sample wells overlaying the gel pieces. Electrophoresis is then carried out until the Bromophenol blue tracking dye reaches the bottom of the stacking gel. It is then halted for 30 min to allow the protease to digest the proteins contained within the gel pieces. As an alternative to temporarily discontinuing electrophoresis, some workers prefer to carry out the stacking and digestion at a very low current. Electrophoresis is then resumed and the separation completed in the normal manner. The resulting peptide maps are then visualized by an appropriate method, such as staining or autoradiography (see Chapter 9). An example of this methodology is shown in *Figure 6.8*. In stained preparations one (or more) of the bands will be due to the protease enzyme.

If a chemical method is employed, the gel pieces have to be subjected to cleavage *in situ* prior to application to the secondary gel. The procedure is, therefore, more time consuming. However, unlike proteolytic digests, chemical cleavages do not contribute to the final peptide map.

References

1. Reynolds, J.A. and Tanford, C. (1970) *J. Biol. Chem.*, **243**, 5161–5165.
2. Shapiro, A.L., Viñuela, E. and Maizel, J.V. (1967) *Biochem. Biophys. Res. Commun.*, **28**, 815–820.
3. Weber, K. and Osborn, M. (1969) *J. Biol. Chem.*, **244**, 4406–4412.
4. Swank, R.T and Munkres, K.D. (1971) *Anal. Biochem.*, **39**, 462–477.
5. Anderson, B.L., Berry, R.W. and Teber, A. (1983) *Anal. Biochem.*, **132**, 365–375.
6. Wyckoff, M., Rodbard, D. and Chrambach, A. (1977) *Anal. Biochem.*, **78**, 459–482.
7. Allen, R.C., Saravis, C.A. and Maurer, H.R. (1984) *Gel Electrophoresis and Isoelectric Focusing of Proteins*. Walter de Gruyter, Berlin.
8. Laemmli, U.K. (1970) *Nature*, **227**, 680–685.
9. Ornstein, L. (1964) *Ann. N.Y. Acad. Sci.*, **121**, 321–349.
10. Davis, B.J. (1964) *Ann. N.Y. Acad. Sci.*, **121**, 404–427.
11. Neville, D.M. (1971) *J. Biol. Chem.*, **246**, 6328–6334.
12. Jovin, T.M. (1973) *Biochemistry*, **12**, 871–879.
13. Lambin, P. (1978) *Anal. Biochem.*, **85**, 114–125.
14. Poduslo, J.F. and Rodbard, D. (1980) *Anal. Biochem.*, **101**, 394–406.
15. Hashimoto, F., Horigome, T., Kanbayashi, M., Yoshida, K. and Sugano, H. (1983) *Anal. Biochem.*, **129**, 192–199.
16. Schagger, H. and von Jagow, G. (1987) *Anal. Biochem.*, **166**, 368–379.
17. Rothe, G.M. and Maurer, W.D. (1986) in *Gel Electrophoresis of Proteins* (M.J. Dunn, ed.). Wright, Bristol, pp. 37–140.
18. Poduslo, J.F. (1981) *Anal. Biochem.*, **114**, 131–139.
19. Traut, R.R., Casiano, C. and Zecherle, N. (1989) in *Protein Function: a Practical Approach* (T.E. Creighton, ed.). IRL Press, Oxford, pp. 101–133.
20. Cleveland, D.W., Fischer, S.G., Kirschner, M.W. and Laemmli, U.K. (1977) *J. Biol. Chem.*, **252**, 1102–1106.
21. Gooderham, K. (1986) in *Gel Electrophoresis of Proteins* (M.J. Dunn, ed.). Wright, Bristol, pp. 312–322.
22. Andrews, A.T. (1990) in *Gel Electrophoresis of Proteins: a Practical Approach* (B.D. Hames, and D. Rickwood, eds). IRL Press, Oxford, pp. 301–319.
23. Rothe, G.M. (1982) *Electrophoresis*, **3**, 255–262.
24. Allore, R.J. and Barber, B.M. (1984) *Anal. Biochem.*, **137**, 523–527.

7 Isoelectric Focusing

7.1 Background

Isoelectric focusing (IEF) is a high resolution method in which proteins are separated in the presence of a continuous pH gradient. Under these conditions, proteins migrate according to their charge properties until they reach the pH values at which they have no net charge (i.e. their isoelectric points, pI). The proteins will, therefore, attain a steady state of zero migration and will be concentrated or focused into narrow zones.

IEF is a very high resolution technique and is routinely able to separate components differing in pI by only 0.02 pH units. If very narrow immobilized pH gradients (see Section 7.3) are used, differences in protein pI of as little as 0.001 pH units can be resolved. This makes IEF a very powerful tool for the analysis of proteins and, in combination with SDS–PAGE, allows characterization of a particular protein sample in terms of both charge and size. Indeed, these two techniques can be combined to form a two-dimensional technique with an almost unique capacity for the analysis of complex protein mixtures in terms of charge and size heterogeneity (see Chapter 8).

IEF is the subject of several excellent review texts, to which the reader is recommended for additional information on this technique [1–6].

7.2 IEF using synthetic carrier ampholytes

7.2.1 Basic principles

The most popular method for generating pH gradients for IEF is the incorporation of low molecular weight amphoteric compounds, known as synthetic carrier ampholytes, into a polyacrylamide gel matrix. These compounds have closely spaced pI values encompassing the pH range of interest. In the absence of an electric field, the ampholytes are

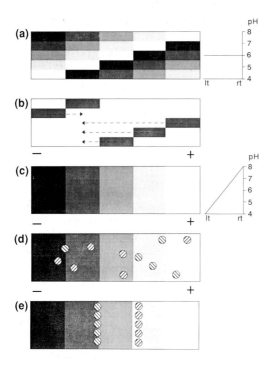

FIGURE 7.1: *Schematic diagram to illustrate the basic principles of isoelectric focusing in a polyacrylamide gel, using a mixture of synthetic carrier ampholytes in the pH 4–8 range. (**a**) In the absence of an electric field the carrier ampholytes are randomly distributed and the pH across the gel is uniform. The different carrier ampholytes are indicated by ▉ pI = 4, ▉ pI = 5, etc. (**b**) When an electric field is applied, each ampholyte, irrespective of its starting position, moves to a position of net zero charge. Only the ▉ species is shown. (**c**) When each ampholyte reaches its isoelectric point, a linear pH gradient is established across the gel. (**d**) Protein molecules (⊘ pI = 5.5, ⊘ pI = 6.7) can be applied at any point in the gel. (**e**) Each protein migrates until it reaches a position in the pH gradient equivalent to its pI value, at which point it carries no net charge and becomes focused into a narrow zone.*

randomly distributed throughout the gel (*Figure 7.1a*). When an electric field is applied, the ampholyte molecules will start to migrate to one or other of the electrodes depending on their net charge (*Figure 7.1b*). For example, the most acidic ampholyte species with the lowest pI will have the highest negative charge and will migrate towards the anode until its net charge is zero, where it concentrates in a narrow zone. In this way each of the ampholyte molecules, irrespective of its starting position, will migrate to a point where its net charge is zero (*Figure 7.1c*). Because each of the ampholytes has a high buffering capacity, the pH of the surrounding medium is equal to its own pI value. Once this process of stacking is complete and a steady state is reached, a continuous pH gradient is formed within the gel (*Figure 7.1c*).

A protein applied to the gel (*Figure 7.1d*), provided that its pI falls within the range of the ampholytes used to generate the pH gradient, will migrate until it reaches a pH equivalent to its own pI (*Figure 7.1e*). Unlike other forms of gel electrophoresis, the sample can be applied anywhere in the gel, as molecules with the same pI will always migrate to the same position. The proteins become highly concentrated or focused into narrow zones. If they diffuse out of the zone, they will become charged and then under the influence of the electric field migrate back to the position of zero net charge.

7.2.2 Gel concentration

As IEF separates proteins on the basis of their charge properties, it is essential to use conditions which minimize molecular sieving effects. This necessitates the use of gels of low acrylamide concentrations in the range 3–5%T. Very large molecules or multimolecular complexes, with M_r values of approximately 10^6, can be separated by IEF using agarose or composite polyacrylamide–agarose gels as the support matrix.

7.2.3 Gel composition

IEF can be performed under native conditions where the gels contain only the added carrier ampholytes. Non-ionic (e.g. Triton X-100) or zwitterionic (e.g. CHAPS) detergents can be added to gels in the range 0.5–5% (w/v) to increase sample solubility during IEF. It can also be carried out under denaturing conditions in the presence of high concentrations of urea (typically 8 M), either in the absence or in the presence of non-ionic or zwitterionic detergents. For gels containing urea, it is essential to minimize the potential of cyanate ions, formed from the breakdown of urea, to carbamylate the sample proteins (see Section 5.2), as this results in serious artifacts in the IEF separation. It should also be noted that the use of highly charged detergents such as SDS is incompatible with IEF, although it is sometimes possible to solubilize a protein sample with SDS and subsequently replace the SDS in the sample with non-ionic or zwitterionic detergent. Thus, samples which require stringent conditions for their solubilization are often not amenable to analysis by IEF. It is also important to minimize the salt concentration of samples (e.g. by gel filtration, ultrafiltration or dialysis) in order to avoid band distortion, extended focusing times and excessive heating effects during IEF.

7.2.4 Choice of ampholytes

Several procedures have been described for the synthesis of carrier ampholytes for IEF. The two main approaches for their synthesis

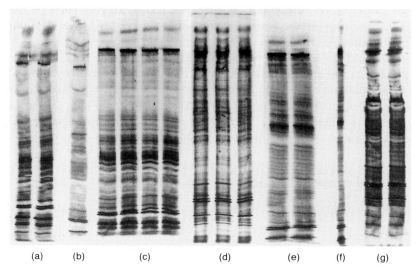

FIGURE 7.2: IEF gels containing 8 M urea and 2% (w/v) NP-40 of [^{35}S]methionine-labeled human skin fibroblast proteins using different synthetic carrier ampholytes. **(a)** 4% (w/v) Pharmalyte 3–10, **(b)** 4% (w/v) Ampholine 3.5–9.5, **(c)** 3% (w/v) Ampholine 3.9–9.5, **(d)** 1.7% (w/v) Pharmalyte 3–10, 1.7% (w/v) Servalyte 2–11, 0.6% (w/v) Pharmalyte 5–8, **(e, f)** 1.28% (w/v) Pharmalyte 3–10, 1.28% (w/v) Servalyte 2–11, 0.44% (w/v) Pharmalyte 5–8, **(g)** 1.0% (w/v) Pharmalyte 3–10, 1.0% (w/v) Servalyte 2–11, 0.24% (w/v) Pharmalyte 5–8, 0.6% (w/v) Ampholine 3.5–9.5, 0.24% (w/v) Ampholine 4–6. Cathode at top. Reproduced from reference [9] with permission from VCH.

involve either reacting different oligoamines (tetra-, penta-, and hexa-amines) with acrylic acid [7], or the co-polymerization of amines, amino acids, and dipeptides with epichlorohydrin [1]. These procedures generate wide-range mixtures containing hundreds, or even thousands, of different amphoteric species with pI values evenly distributed across the pH range. These carrier ampholyte preparations can cover the range pH 2.5–11, but in practice IEF is limited to the range pH 3.5–10, since very few proteins have pI values outside of this range. Indeed, the pI values for most proteins are clustered in the pH 4–6 region [8].

Carrier ampholyte preparations covering the range pH 2.5–11 are available commercially (see Appendix B), for example Pharmalyte (Pharmacia–LKB), Ampholine (Pharmacia–LKB), Servalyte (Serva), Resolyte (Merck), and Biolyte (Bio-Rad). As a result of the variety of synthetic procedures used, different commercial carrier ampholyte preparations are often found to give better resolution in different pH regions of IEF gels. For example, we have found [9] that for human skin fibroblast proteins, Ampholine yielded superior cathodic resolution, Pharmalyte gave better separation in the mid-range, and Servalyte gave the best resolution of acidic proteins (*Figure 7.2*). It is possible by blending these different full-range ampholyte preparations to generate a mixture containing a greater diversity of charge varieties and so incorporate the advantages of each preparation and thus maximize separation and resolution (*Figure 7.2*).

In addition to full-range ampholytes, narrow range preparations are available to produce gradients spanning as little as 1 pH unit. Such shallow gradients are capable of resolving proteins differing in pI by as little as 0.01 pH units. These narrow range preparations can also be added to a wide-range ampholyte mixture to flatten the slope of the pH gradient in the region where the proteins of interest, in a complex mixture of proteins, are distributed.

No simple guide can be given to the choice of the pH gradient interval, particular ampholyte preparation or blended mixture best suited for the separation of any particular protein sample. A wide-range pH gradient is indicated for complex mixtures containing proteins with a wide range of pI values. If a mixture of ampholytes is to be used, the best blend can only be established empirically to optimize the separation of the particular protein sample. If the pI values of the proteins of interest are already known, or have been established by IEF using a wide-range pH gradient, then an appropriate narrow-range ampholyte mixture can be used to optimize the separation.

7.2.5 Gel preparation and apparatus

IEF techniques are best performed using slab gels run on a horizontal flat-bed apparatus (see Section 3.7.2) with an efficient cooling platen to cope with the Joule heating caused by the high field strengths generally employed (typically around 1000 V). It is also beneficial to use thin (0.5 mm) or ultrathin (down to 20 μm) gels, which can be cast on thin plastic supports (e.g. GelBond PAG^{TM} film) to improve heat dissipation and ease of handling.

Gels are cast, preferably on plastic support films, using the simple cassette procedure described in Section 3.7.2. Sample wells should not be created within the gel itself as this perturbs the electric field and, consequently, the pH gradient during IEF. Samples should be applied using silicone rubber applicator strips placed on the surface of the gel once it has been set up on the platen of the flat-bed apparatus. These strips remain on the gel for the duration of the IEF run.

A range of ready-made, wet IEF gels bound to plastic supports is commercially available (Ampholine PAGplates from Pharmacia–LKB), covering both broad (pH 3.5–9.5) and narrow (pH 4–5, 4–6.5, and 5.5–8.5) pH intervals. These ready-made gels are simple to use and give reproducible patterns. However, they are relatively expensive and restricted to use under native conditions as it is difficult to incorporate detergents and/or urea into the system.

7.2.6 Running conditions

Non-volatile free acids and bases are used as electrolytes for the anode (anolyte) and cathode (catholyte) for IEF. In horizontal systems, filter paper strips impregnated with the electrolytes are applied to the surface of the gel. As the electrolyte volume is limited using this technique, quite strong solutions are used. For most wide-range pH gradients, solutions of 1 M phosphoric acid and 1 M sodium hydroxide can be used as anolyte and catholyte, respectively. For narrow-range pH gradients it is better if electrolytes are chosen to match the range of the ampholyte mixture being used. Mixtures of carrier ampholytes are often recommended as electrolytes for narrow-range pH gradients.

The IEF run should be carried out with the cooling system set to 2–4°C. However, if urea is present within the gel, a higher temperature setting of 10–15°C must be used to avoid precipitation of the urea.

In theory, samples can be applied at any position on the gel, but in practice the optimal application point should be determined for each particular type of sample by trial and error. Samples should be applied away from their pI (where proteins exhibit their minimum solubility), and in practice, most samples are applied near the cathode.

Lower voltages must be used at the start of the IEF run as the initial current will be relatively high. As the pH gradient becomes established, the current falls and the applied voltage can be progressively increased. The use of higher voltages produces sharper protein zones. It is, therefore, best to use a power pack capable of providing constant applied power conditions for IEF (see Appendix B). Typical running conditions for a pH 3.5–10 IEF gel are given in *Table 7.1*.

The optimal run time used for IEF should also be determined for each type of sample to be analyzed. Different proteins can focus at different speeds and gel additives such as urea or glycerol can extend the time required for focusing to be achieved. In theory, IEF is an equilibrium steady-state technique, so that once the individual sample proteins have attained positions in the gel equivalent to their pI values, a stable band pattern should be produced. Thus, to ensure that equilibrium focusing of a particular sample has been achieved, it is recommended that the sample be applied at various positions across the pH gradient, to determine the focusing time required to obtain a constant separation pattern, independent of where the sample was applied.

ISOELECTRIC FOCUSING

TABLE 7.1: *Suggested running conditions for native IEF at 10°C using synthetic carrier ampholytes in the pH 3.5–9.5 range*

Gel thickness (mm)	0.1	0.2	0.3	0.4	0.5
Pre-run					
Voltage (V)	max	max	max	max	max
Current (mA)	10	10	15	15	15
Power (W)	10	10	15	15	15
Time (min)	10	20	20	20	20
Sample applied: focusing					
Voltage (V)	max	max	max	max	max
Current (mA)	50	50	50	50	50
Power (W)	15	15	20	20	25
Time (min)	45	60	60	60	60
Total time (min)	55	80	80	80	80

The gels should be run at the maximum voltage setting of the power pack available (1000–3000 V). These conditions are for a 26 cm wide IEF gel with a 10 cm separation length. Smaller gels should be run under reduced power and current settings; for example, if only half a gel is used, reduce the power and current settings to one half. Sharper bands can be obtained by increasing the power setting by 5 W for the last 15 min of the run

However, due to limitations inherent in the use of synthetic carrier ampholytes (see Section 7.2.8), pH gradients achieved by IEF in polyacrylamide gels are not infinitely stable.

The run time required to obtain the best separation of the particular sample must be determined empirically. As this is dependent on the applied voltage, it is customary to express the running conditions as the product of applied voltage and time, that is volthours (Vh). The separation length of the IEF gel is another factor which determines the time required for IEF. At a given applied voltage, as the separation length of the gel is increased, the field strength (volts/cm) is reduced. Moreover, the proteins also have further to travel to reach their respective pI values. The run time is then dependent on the square of the separation distance, l. Some investigators, therefore, prefer to express IEF running conditions as Vh/l^2.

7.2.7 Estimation of pH gradient

After the IEF run is complete, the pH gradient in the gel can be measured most conveniently using a flat membrane surface pH electrode. Alternatively, a narrow strip can be cut across the focusing direction of the gel. This strip can then be divided into segments which are eluted into a small volume of 10 mM KCl, previously degassed to remove dissolved CO_2. If the IEF gel contains urea, then this solution should also contain urea at the same concentration as that present in the gel. The pH of the eluted samples can then be measured at a defined temperature and the appropriate compensatory factor applied to

TABLE 7.2: *Compensation factors for the effects of temperature and 1 M urea on the determination of operational pH during IEF*

Effect of temperature (25–4°C)		Effect of 1 M urea	
pH	pH shift	pH	pH shift
—	—	3.0	+0.09
3.5	+0.00	3.5	+0.08
4.0	+0.06	4.0	+0.08
4.5	+0.10	4.5	+0.07
5.0	+0.14	5.0	+0.06
5.5	+0.20	5.5	+0.06
6.0	+0.23	6.0	+0.05
6.5	+0.28	6.5	+0.05
7.0	+0.36	7.0	+0.05
7.5	+0.39	7.5	+0.06
8.0	+0.45	8.0	+0.06
8.5	+0.52	8.5	+0.06
9.0	+0.53	9.0	+0.06
9.5	+0.54	9.5	+0.05
10.0	+0.55	—	—

Reproduced from reference [3] with permission from Oxford University Press.

determine the pH value operative at the temperature at which the IEF run was carried out. The effects of temperature and urea on pH are illustrated in *Table 7.2*.

An alternative approach is to use a set of standard marker proteins of known pI values to calibrate the pH gradient achieved during IEF. Several kits of such marker proteins, covering various pH ranges, are commercially available (see Appendix B). Some of these proteins are colored which means that the progression of the IEF run can be monitored continuously. One problem with these kits is that multiple protein bands are observed when sensitive staining protocols are used

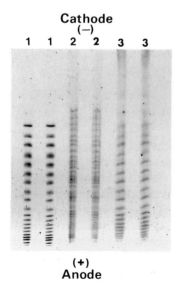

FIGURE 7.3: *Horizontal flat-bed IEF gel of carbamylated pI standards for calibration of the pH gradient. Focusing conditions: 2000 Vh using a synthetic carrier ampholyte mixture (Ampholine, pH 3.5–10, and Resolyte, pH 4–8, 1:1) in the presence of 8 M urea. Lane 1, carbonic anhydrase; lane 2, creatine phosphokinase; lane 3, glyceraldehyde-3-phosphate dehydrogenase.*

(e.g. silver, see Section 9.6). These IEF marker kits work well under native conditions, but many of the standard proteins used contain multiple subunits. Thus, under denaturing conditions, such as in the presence of urea and detergents, these proteins are dissociated into their constituent subunits and form multiple protein zones on IEF. The pI values of the individual subunits is generally not known.

Another approach for pH gradient determination is to use markers based on the use of multiple charged forms of a particular protein. This is achieved by heating the protein in the presence of urea under conditions which ensure the generation of carbamylated derivatives. Under appropriate conditions, the multiple carbamylated forms of the protein produce a continuous series ('train') of bands, each of which differs from the previous form by the addition of one negative charge. Examples of this approach are shown in *Figure 7.3*. Standards of this type are applicable both to native and denaturing IEF gels. The disadvantage of these standards is that the pH gradient cannot be calibrated in terms of absolute pH, but only in terms of relative charge changes from the unmodified protein. A conversion table for the charge train standards produced by the carbamylation of creatine kinase has been produced by Merck (*Table 7.3*).

TABLE 7.3: *Relative pI values of carbamylated creatine kinase (CK) markers*

Carbamylated CK band number	Relative pI	Carbamylated CK band number	Relative pI
1	7.0	20	5.88
2	6.94	21	5.80
3	6.91	22	5.73
4	6.89	23	5.64
5	6.86	24	5.57
6	6.81	25	5.51
7	6.76	26	5.46
8	6.71	27	5.41
9	6.58	28	5.39
10	6.51	29	5.36
11	6.44	30	5.33
12	6.37	31	5.31
13	6.31	32	5.29
14	6.27	33	5.26
15	6.22	34	5.15
16	6.14	35	5.08
17	6.09	36	4.95
18	6.01	37	4.95
19	5.94	38	4.95

Data reproduced with permission from Merck Ltd.

7.2.8 Limitations of IEF using synthetic carrier ampholytes

The main disadvantages of this type of IEF are concerned largely with the properties of the synthetic carrier ampholytes used to generate the pH gradients. In theory, IEF is a stable, equilibrium technique, but in practice the electroendosmotic properties of IEF gels result in long-term instability of pH gradients. This phenomenon results in cathodic drift, so that with time the pH gradient, together with the separated proteins within it, migrates towards the cathode, thus causing decay of the pH gradient and loss of protein from the gel. Anodic drift can also occur under certain circumstances.

The effects of cathodic drift can be minimized, but not totally overcome, by two alternative approaches. The first approach is to increase the pH of the anolyte solution (i.e. decreasing the proton molarity at this electrode) [1]. The second approach is to employ 'balanced matrices' involving the use of acrylamide derivatives, such as 3-dimethylamino propylmethacrylamide (DMAPMA) or methacrylamido propyltrimethyl ammonium chloride (MAPTAC), which contain tertiary or quaternary amines [1]. These groups act to balance the inherent negative charge present on polyacrylamide matrices due to: (i) trace impurities of acrylic acid in the gel, (ii) covalent incorporation of catalysts into polyacrylamide chains, and (iii) hydrolysis of amide groups to acrylic acid in the basic region of the gel.

Another major problem with IEF is that there is considerable batch to batch variation in ampholyte preparations, due to the complex procedures involved in their synthesis. This results in variations in pH gradients occurring during IEF. It is, therefore, important to perform quality testing on different batches of ampholytes to ensure that reproducible protein separations are obtained. In addition, it is difficult to generate particular pH gradient shapes optimized to separate particular proteins (i.e. pH gradient engineering) other than by trial and error.

Finally, it is important to realize that there are classes of compounds, generally highly charged species such as heparin, polyanions, polycations and certain dyes, which have been shown to display artifactual heterogeneity when separated by IEF. This effect has been ascribed to interaction of these compounds with the synthetic carrier ampholytes themselves, resulting in the formation of a complex with an altered pI value.

7.3 IEF using immobilized pH gradients

7.3.1 Basic principles

The limitations inherent in the use of synthetic carrier ampholytes for the generation of pH gradients for IEF (see Section 7.2.8), prompted the development of Immobiline reagents (Pharmacia–LKB) for the preparation of polyacrylamide gels containing immobilized pH gradients (IPG). The development of these reagents was the result of a collaborative effort between Dr B. Bjellqvist's team at LKB (now Pharmacia–LKB Biotechnology), and the research groups of Dr A. Görg (Munich) and Dr P.G. Righetti (Milan) [10]. A detailed review of IPG technology can be found in the monograph devoted to this topic by Righetti [6].

The Immobiline reagents are a series of seven acrylamide derivatives each with the general structure $CH_2=CH–CO–NH–R$, where R contains either a carboxyl (acidic Immobilines) or a tertiary amino (basic Immobilines) group, forming a series of buffers with different pK values distributed throughout the pH 3–10 range. The pK values of the Immobiline reagents currently obtainable commercially from Pharmacia–LKB are: 3.6, 4.4, 4.6, 6.2, 7.0, 8.5, and 9.3. Additional reagents have also been described for generating extended pH gradients, such as quaternary amino ethyl acrylamide, pK > 12.0, and 2-acrylamido-2-methyl propane sulfonic acid, pK 1.0 [11]. These additional reagents have recently become available from Polysciences (pK 1.0), IBF (pK > 12) and Fluka (pK 1.0, 10.3 and 12.0). If Immobilines of the appropriate pK values are available, narrow pH gradients may be generated by incorporating inverse gradients of Immobilines into a polyacrylamide gel during polymerization. In other cases, and particularly for external pH gradients, more complex mixtures of Immobilines may be required (see Sections 7.3.2 and 7.3.4). The buffering groups forming the pH gradient are covalently attached via vinyl bonds to the polyacrylamide backbone, thereby effectively immobilizing the pH gradient within the gel.

Although electroendosmosis is not affected, the immobilized pH gradient is not subject to cathodic drift, so that it should be infinitely stable, allowing true equilibrium IEF to be performed. However, in practice, electroendosmosis can still cause problems during IPG IEF in the extreme pH ranges (i.e. below pH 5.0 and above pH 9.0). Nevertheless, the stability of the pH gradients combined with the defined chemical nature of the Immobiline reagents should ensure greater reproducibility of protein separations than can be achieved using conventional IEF based on the use of synthetic carrier ampholytes.

7.3.2 pH gradients

Extensive recipes are available for a wide variety of pH gradients that can be generated using the Immobiline reagents. These are available from the manufacturers (Pharmacia–LKB) and are provided with the reagents. In addition, many additional recipes have been published [3,6] and sophisticated computer programs (see Appendix B) have been developed which can formulate the appropriate conditions to generate a pH gradient of any defined range and shape [1].

(a) Narrow and ultra-narrow pH gradients. It is possible to generate narrow and ultra-narrow IPGs spanning from as little as 0.1–1 pH units. Such narrow and ultra-narrow pH gradients have an extremely high resolving power and are claimed to be able to resolve proteins with a difference in pI of as little as 0.001 pH units. An example of the use of such narrow pH gradients is illustrated in *Figure 7.4*.

(b) Extended pH gradients. Extended pH gradients (>1 pH unit) can also be readily generated using the Immobiline reagents. With the standard set of reagents it is possible to generate pH gradients spanning the pH 3–10 interval, but greater flexibility is gained if additional reagents are used (see Section 7.3.1). Using this approach pH gradients spanning the interval pH 2.5–11 have been generated and successfully used for IEF [12].

(c) Non-linear pH gradients. The majority of published recipes for generating IPGs are designed to produce rigorously linear pH gradients. However, it can often be desirable to use pH gradients which are flattened in particular regions where many proteins are located. This concept of pH gradient engineering is relatively easy to accomplish reproducibly and controllably using IPG technology, particularly if a microcomputer controlled system (see Section 7.3.4) is available for gel preparation.

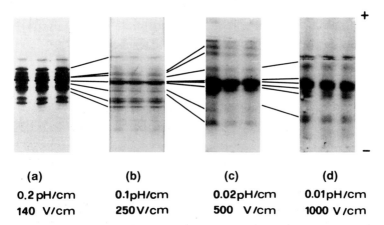

(a)	(b)	(c)	(d)
0.2 pH/cm	0.1 pH/cm	0.02 pH/cm	0.01 pH/cm
140 V/cm	250 V/cm	500 V/cm	1000 V/cm

FIGURE 7.4: Resolving power of IPG IEF. Ovalbumin focused *(a)* using synthetic carrier ampholyte IEF (Ampholine, pH 4–6), and using IPG with varying pH slopes *(b–d)*. Reproduced from reference [18] with permission from Elsevier Science Publishers BV.

(d) Hybrid or mixed-bed IEF. Some problems have been experienced with the use of wide-range IPGs, including slow entry of sample proteins, lateral band spreading, prolonged focusing times, and increased electroendosmosis. These problems have been attributed either to the inherently low conductivity of the Immobiline system [13] or to the IPG matrix being rather hydrophobic [3]. Using a technique known as hybrid or mixed-bed IEF, improved separations can sometimes be obtained by the addition of low concentrations (typically 0.5%, w/v) of synthetic carrier ampholytes of the appropriate range to the IPG IEF gels. Carrier ampholyte concentrations in excess of 1% (w/v) should be avoided as the pH gradient can be disrupted and excessive water exudation from the gel can occur.

7.3.3 Problems of Immobiline reagents

Although IPG IEF has many advantages over conventional IEF using synthetic carrier ampholytes, many investigators have encountered problems when attempting to convert to this technology. It has now been established that many of these problems were associated with poor stability and limited shelf life of the reagents. Degradation of the chemicals occurs both through autopolymerization and by hydrolysis of the amide bond. The latter event results in the release of acrylic acid and either an amino acid (acidic Immobilines) or a diamine (basic Immobilines). This degradation occurs, albeit slowly, even if the reagents are aliquoted and stored at $-20°C$. These problems of reagent stability have now been overcome and a new set of reagents known as Immobiline II are available from Pharmacia–LKB. The chemicals are now supplied as 0.2 M solutions, dissolved in water containing an inhibitor of polymerization and hydrolysis (hydroquinone monomethylether) for the acidic species and *n*-propanol (propan-1-ol) for the basic reagents.

7.3.4 Gel preparation

IPG IEF gels are prepared, normally as thin (typically 0.5 mm) 3 or 4%T slab gels on plastic supports, by generating a gradient using the appropriate Immobiline solutions to construct the desired pH gradient (see Section 7.3.2). IPG gels are cast in the same way as conventional gradient polyacrylamide gels (see Section 3.8.2), with a density gradient to stabilize the Immobiline concentration gradients, using either a simple two-vessel gradient mixer (*Figure 7.5*) or a more sophisticated microcomputer controlled system (*Figure 7.6*). The latter has the advantage that more reproducible IPG IEF gels can be produced and pH gradient engineering can be more readily accomplished. In order to achieve optimal and reproducible polymerization, it is recommended that the gel is heated at 50°C for 1 h.

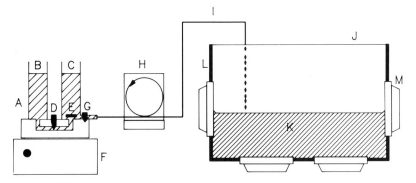

FIGURE 7.5: Simple system for the preparation of IPG gels. A, Two-chamber gradient mixer; B, reservoir containing light solution; C, mixing chamber containing dense solution; D, valve; E, stirring bar; F, magnetic stirrer; G, stopcock; H, peristaltic pump; I, tubing; J, gel cassette; K, gel solution; L, gasket; M, clamps.

After polymerization is complete, the gel is removed from its casting cassette and must be washed thoroughly in distilled water to remove any remaining catalysts and unreacted monomers and Immobilines. Unfortunately, the gel takes up water during this step. This excess water can be removed using a flow of unheated air until the gel returns to its original weight. However, it is generally more convenient to dry the gel completely, as dried gels can be stored at $-20°C$ for extended periods. A range of ready-made, dry IPG IEF gels with different pH gradients (Immobiline DryPlates) can be obtained from Pharmacia–LKB. Unfortunately, the widest ready-made IPG gels currently available span pH 4–7 and no alkaline (>pH 7) ranges are available. Hopefully, the range of these products will be extended in the future as

FIGURE 7.6: Computer-controlled stepmotor-driven buret system for preparing IPG IEF gels. 1, Computer; 2, floppy disk drive; 3, printer; 4, four burets; 5, mixing chambers; 6, magnetic stirrer; 7, gel cassettes. Reproduced from reference [19] with permission from VCH.

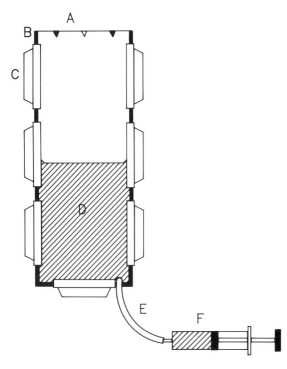

FIGURE 7.7: Re-swelling cassette for dry IPG gels. The dried IPG gel, on its plastic support, is inserted into the cassette, which is filled from the bottom, using a syringe, with the appropriate re-swelling solution. A, Gel cassette; B, gasket; C, clamps; D, gel solution; E, tubing; F, syringe.

they are an excellent way of establishing IPG IEF techniques and they avoid the need for the rather complicated gel preparation involved in the production of custom-made IPG gels.

When required for an IEF experiment, the commercial or home-made dried gels are re-swollen in the desired solution. Often the gels are simply re-swollen in water; this system also provides an extremely effective way of introducing additives such as urea, non-ionic or zwitterionic detergents, and synthetic carrier ampholytes (for hybrid IEF) into the gel matrix. The best method of re-swelling dried IPG gels is to use a cassette similar to the one in which the gel was cast, but using a spacer gasket which is thicker to take into account the thickness of the plastic support on which the gel has been prepared. The cassette is then filled with the appropriate solution (or water), preferably from the bottom to avoid bubble formation, and the gel allowed to re-swell (*Figure 7.7*). Re-swelling takes 1–2 h when distilled water is used, but extended times are required (18 h) if urea and detergents are present in the re-swelling solution.

7.3.5 Sample application

Very small samples can be applied directly to the surface of the IPG gel in the form of a small droplet or streak. However, as in the case of conventional IEF using synthetic carrier ampholytes, the best method of sample application is to use a silicone rubber applicator strip containing wells to hold the sample. It is generally recommended that IPG gels should not be pre-run prior to sample application; pre-running makes sample entry less efficient and causes lateral spreading of protein zones.

Although the IPG technique is more tolerant than conventional IEF to the presence of salt in the sample, it is recommended that this be kept below 40 mM, or problems of sample modification and/or precipitation can occur. Salts of strong acids and bases (e.g. NaCl, Na_2SO_4, Na_2HPO_4) should be avoided, so that if it is essential to include salt in the sample it is best to use those formed from weak acids and bases (e.g. Tris–acetate, Tris–glycinate). It is also beneficial to add carrier ampholytes to samples containing high levels of salt (e.g. 10% (w/v) to 100 mM salt).

IPGs are also considerably more tolerant of high sample protein loads than conventional IEF. Righetti claims a ten times higher loading capacity for IPGs, making them an ideal tool for certain preparative applications [6].

It should be noted that certain classes of proteins, including histones, high-mobility group chromatin proteins, albumin, and ferritin can interact with IPG matrices and in some cases can form insoluble complexes [14]. This probably involves both ionic and hydrophobic interactions, and there is some evidence that this may be a more general problem leading to protein loss and increased background staining in IPG IEF gels [15].

7.3.6 Running conditions

Glutamic acid (10 mM) and lysine (10 mM) are commonly used as the anolyte and catholyte, respectively, for IPG IEF gels. Alternatively, carrier ampholytes of the same or of a narrower pH range than the IPG can be used. For mixed-bed gels or for samples with relatively high salt concentrations, it is often satisfactory to use distilled water for both electrolytes.

As in the case of conventional IEF, power packs capable of running gels under conditions of constant power, in addition to constant voltage or current, are the best for use in IPG techniques. An initial low voltage of about 500 V (15 mA, 20 W limiting) for 1–2 h should be used to allow slow sample entry into the gel. This should be extended to 4 h if the sample contains high levels of salt. The gels should then be run at 2000–3000 V (5 mA, 5 W limiting) for 3–5 h in the case of broad pH gradients. Longer focusing times of about 18 h are required for pH gradients spanning 1 pH unit or less. Some investigators have successfully used higher field strengths (5000–10 000 V) for IPG IEF, but such high voltages cannot be used for mixed-bed IEF gels. In any case, a short period (e.g. 1 h) at 5000 V can be used at the end of the run to enhance band sharpness further.

As discussed in Section 7.2.6, the run time required to obtain the best separation of the particular sample being analyzed must be determined empirically and the running conditions should be expressed in Vh (or Vh/l^2).

7.3.7 Estimation of pH gradient

It is impossible to measure pH values in IPG IEF gels using a surface pH electrode or by eluting slices of the gel (see Section 7.2.7) since there are no water-soluble ampholytes present. For narrow pH gradients of 1 pH unit or less, the best method is to interpolate the pI value of a particular protein band by its position on the gel, assuming a linear pH gradient between the two extremities. For wider pH gradients it is best to use pI marker kits where these are available (see Section 7.2.7).

7.4 Titration curves

The two-dimensional method described in this section results in band patterns which represent the titration curves for the proteins present in the sample. A polyacrylamide slab gel containing synthetic carrier ampholytes is cast with a trench in the middle (*Figure 7.8*). The gel is focused, in the absence of a sample, to establish a pH gradient within the gel. At this stage, the electrode strips and underlying areas of gel must be cut away (*Figure 7.8*). If this is not done, these regions which contain strong acid and base generate excessive heat in the second dimension. The prefocused gel is then rotated through 90°. New electrodes are applied and the trench is filled with the sample to be analyzed (*Figure 7.8*). An electric field is then applied perpendicular to the pH gradient (*Figure 7.8*). This technique is reviewed in detail in reference [1] and some of its applications are briefly described in the following sections.

7.4.1 Characterization of genetic mutants

Differential titration curves can be performed by running a mixture of a protein and its genetic mutants. The shape of the respective titration

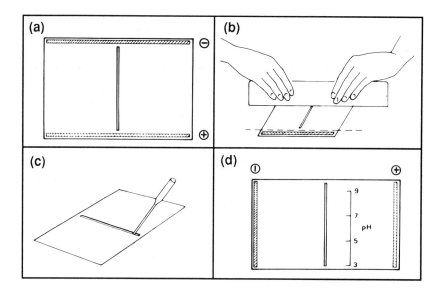

FIGURE 7.8: *Experimental procedure for generating titration curves. (a) The pH gradient is first formed by focusing the synthetic carrier ampholyte mixture; (b) the electrode strips are removed; (c) The sample is applied in a trench cut perpendicular to the pH gradient; (d) the second dimension run is started perpendicular to the first dimension axis. Reproduced from reference [1] with permission from Elsevier Science Publishers BV.*

curves then reveals which charged amino acid is substituted in the mutant phenotype. A theoretical application of this technique to variants of hemoglobin in which charged amino acids are replaced by other amino acids (or vice versa) is shown in *Figure 7.9*. In the case of lysine (Lys) mutants, the two curves meet around pH 11, while for mutants with substitutions involving acidic amino acids, aspartic (Asp), and glutamic (Glu) acids, the confluence point is around pH 3. Double charge mutants (e.g. Lys → Glu) or substitutions involving amino acids with the same charge (e.g. Arg → His) can also be detected. These theoretical titration curves have been confirmed experimentally using normal human adult hemoglobin mixed with a variety of genetic mutant forms [16].

7.4.2 Analysis of macromolecular interactions

Using titration curves, it is possible to study the interaction of proteins with low molecular mass ligands and with proteins or other macromolecules. It is also possible to determine dissociation constants (K_d) of ligands to proteins and their pH dependence. The ligand is immobilized in the gel matrix either by simple entrapment, if it is a macromolecule, or by covalent attachment to the gel fibers, if it is a

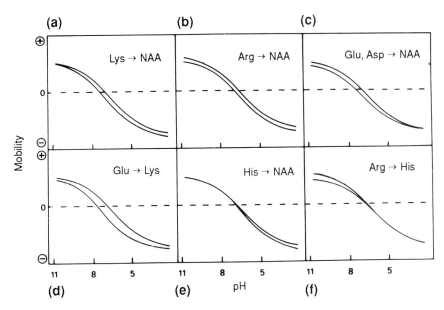

FIGURE 7.9: Theoretical titration curves for (**a**) substitution of a lysine residue by a neutral amino acid, (**b**) substitution of an arginine residue by a neutral amino acid, (**c**) substitution of a glutamic or aspartic acid residue by a neutral amino acid, (**d**) a double charge difference resulting from the substitution of a lysine residue by a glutamic or aspartic acid residue, or vice versa, (**e**) substitution of a histidine by a neutral amino acid, and (**f**) substitution of an arginine by a histidine residue. The dotted line represents the zero-mobility plane of the protein, which in practice is the sample application trough. Reproduced from reference [1] with permission from Elsevier Science Publishers BV.

small molecule. In the presence of increasing concentrations of ligand the titration curve of the protein is progressively retarded, in a pH-dependent fashion, and the mobility decrements, when plotted against the ligand molarity in the gel, can be used to calculate K_d values at any pH value. An example of such affino-titration curves is shown in *Figure 7.10* for the binding of the lectin, *Ricinus communis*, to allyl-α-D-galactose co-polymerized in the gel matrix [17]. K_d values can be determined at any pH and then plotted against pH in order to evaluate the pH dependence of K_d (*Figure 7.11*).

7.4.3 Estimation of pK values

A mathematical method has been established by Righetti and his colleagues for the direct pK determination of ionizable groups from the shape of pH–mobility curves [1]. For non-amphoteric ions, a direct determination of either pK_c or pK_a can be made by measuring the pH ($pH_{½}$) corresponding to the ½ mobility in the cathodic or ionic

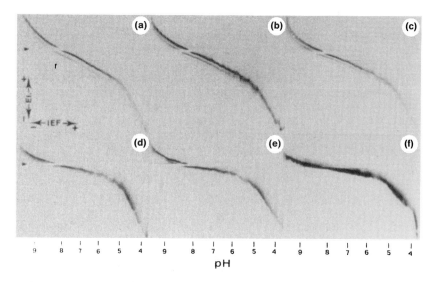

FIGURE 7.10: *Affino-titration curves of the lectin from* Ricinus communis *seeds. The amount of allyl-α-D-galactose co-polymerized in the gel matrix was:* **(a)** *control, no ligand,* **(b)** 1×10^{-5} M, **(c)** 4×10^{-5} M, **(d)** 5×10^{-5} M, **(e)** 7×10^{-5} M, *and* **(f)** 10^{-4} M. *In each case, 200 μg protein was loaded. The arrow heads (a, d) indicate the sample application trench (zero mobility plane). EL = electrophoresis. Reproduced from reference [1] with permission from Elsevier Science Publishers BV.*

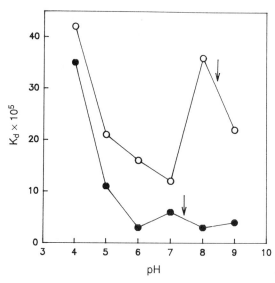

FIGURE 7.11: *Variation of the dissociation constants* (K_d) *as a function of pH for* Ricinus communis *(lower line) and for* Lens culinaris *(upper line) lectins. The* K_d *values were calculated from the retarded mobilities, at different pH values, in affino-titration curve analysis, such as shown in* Figure 7.10. *The two arrows indicate the pI of each lectin. Reproduced from reference [1] with permission from Elsevier Science Publishers BV.*

direction, respectively. For amphoteric species, the $pH_{1/2}$ values must be corrected by a term accounting for the influence of the degree of ionization of the opposite charge on the mobility curve of the ion being measured.

References

1. Righetti, P.G. (1983) *Isoelectric Focusing: Theory, Methodology and Applications*. Elsevier, Amsterdam.
2. Andrews, A.T. (1986) *Electrophoresis: Theory, Techniques and Biochemical and Clinical Applications*. Clarendon Press, Oxford.
3. Righetti, P.G., Gianazza, E., Gelfi, C. and Chiari, M. (1990) in *Gel Electrophoresis of Proteins: a Practical Approach* (B.D. Hames and D. Rickwood, eds). IRL Press, Oxford, pp. 149–216.
4. Righetti, P.G., Gelfi, C. and Gianazza, E. (1986) in *Gel Electrophoresis of Proteins* (M.J. Dunn, ed.). Wright, Bristol, pp. 141–202.
5. Allen, R.C., Saravis, C.A. and Maurer, H.R. (1984) *Gel Electrophoresis and Isoelectric Focusing of Proteins*. Walter de Gruyter, Berlin.
6. Righetti, P.G. (1990) *Immobilized pH Gradients: Theory and Methodology*. Elsevier, Amsterdam.
7. Vesterberg, O. (1969) *Acta Chem. Scand.*, **23**, 2653–2666.
8. Gianazza, E. and Righetti, P.G. (1980) *J. Chromatogr.*, **193**, 1–8.
9. Burghes, A.H.M., Dunn, M.J. and Dubowitz, V. (1982) *Electrophoresis*, **3**, 354–363.
10. Bjellqvist, B., Ek, K., Righetti, P.G., Gianazza, E., Görg, A., Postel, W. and Westermeier, R. (1982) *J. Biochem. Biophys. Meth.*, **6**, 317–339.
11. Chiari, M. and Righetti, P.G. (1992) *Electrophoresis*, **13**, 187–191.
12. Sinha, P., Köttgen, E., Westermeier, R. and Righetti, P.G. (1992) *Electrophoresis*, **13**, 210–214.
13. Altland, K. and Rossman, U. (1985) *Electrophoresis*, **6**, 314–325.
14. Righetti, P.G., Gelfi, C., Bossi, M.L. and Boschetti, E. (1987) *Electrophoresis*, **8**, 62–70.
15. Altland, K., von Eckardstein, A., Banzhoff, A., Wagner, M., Rossman, U., Hackler, R. and Beder, P. (1987) *Electrophoresis*, **8**, 52–62.
16. Righetti, P.G. and Gianazza, E. (1978) *Protides Biol. Fluids*, **27**, 711–712.
17. Ek, K., Gianazza, E. and Righetti, P.G. (1980) *Biochim. Biophys. Acta*, **626**, 356–365.
18. Righetti, P.G., Gianazza, E. and Bjellqvist, B. (1983) *J. Biochem. Biophys. Meth.*, **8**, 89–108.
19. Altland, K. and Altland, A. (1984) *Electrophoresis*, **5**, 143–147.

8 Two-Dimensional Gel Electrophoresis

8.1 Background

Any of the one-dimensional electrophoretic techniques discussed so far will separate complex protein samples and, under optimal conditions, allow the resolution of around 100 distinct zones. In recent years, however, it has become necessary to analyze samples of much higher complexity, such as those represented by whole cells or tissues. An example of this complexity is the human genome which consists of 3×10^9 base pairs of DNA distributed in 23 distinct chromosomes. It has been estimated that there are about 50 000–100 000 genes in the human genome, with perhaps around 5000 of these genes being expressed as proteins in any one distinct cell type. This potential level of complexity is, therefore, not amenable to analysis by any single one-dimensional electrophoretic procedure.

Another serious limitation of one-dimensional electrophoretic methods is that, using any particular procedure, proteins are separated on the basis of only one of their physico-chemical properties (i.e. size, charge, or hydrophobicity). Consequently, the discrete bands which are detected after electrophoresis do not necessarily represent homogeneous proteins. Such zones will contain all proteins in the sample with the same properties which have been exploited for the separation (i.e. charge, mobility (see Section 8.4.2), size).

These two factors have resulted in the search for electrophoretic methods with the potential to separate very complex samples containing several thousands of proteins and to resolve proteins sharing similar physico-chemical properties. The best approach to this problem which is currently available, is to combine two different one-dimensional electrophoretic techniques into a two-dimensional procedure. Ideally the methods used for each dimension should be chosen so that they separate proteins according to different, independent, physico-chemical properties. Thus, if two such methods are selected, which when used alone are capable of separating 100 protein zones, it would be anticipated that theoretically up to 10 000 proteins could be resolved. In practice, this level of resolution has rarely been achieved (see Section 8.11); nevertheless, current methods of two-dimensional polyacrylamide gel electrophoresis (2-D PAGE) are applicable to the analysis of protein expression in whole cells, tissues and even organisms.

The first 2-D electrophoretic separation of proteins can probably be ascribed to Smithies and Poulik who, in 1956, described a 2-D combination of paper and starch gel electrophoresis for the separation of serum proteins [1]. Subsequent developments in electrophoretic technology, such as the use of polyacrylamide as a support medium, the development of discontinuous buffer systems and the use of polyacrylamide concentration gradients were rapidly applied to 2-D separations [2]. In particular, the application of isoelectric focusing (IEF) techniques to 2-D separations made it possible for the first dimension to be based on charge. The coupling of IEF with SDS–PAGE in the second dimension resulted in a 2-D method that separated proteins according to two independent parameters (i.e. charge and size). This methodology was then adapted to a wide range of samples with differing solubility properties by the use of urea and non-ionic detergents for IEF. Thus, by 1975, a 2-D PAGE system had been developed which could be applied to the analysis of protein mixtures from whole cells and tissues [3–5].

In that same year, building on the basis of these developments, O'Farrell described a method of 2-D PAGE optimized for the separation of the proteins of *Escherichia coli* [6]. This method used a combination of IEF in cylindrical 4%T, 5%C gels containing 8 M urea and 2% (w/v) Nonidet P-40 (NP-40) with the discontinuous SDS–PAGE system of Laemmli [7]. Gradient gels were also used for the SDS–PAGE dimension to obtain maximum separation of *E. coli* proteins and sensitive detection was achieved using autoradiography. Using this technique, about 500 polypeptides from *E. coli* were resolved.

The technique described by O'Farrell [6], which is described in subsequent sections of this chapter, has formed the basis for most developments in 2-D PAGE since 1975 and the popularity of this methodology can be judged from the several thousand papers which have been published using this methodology.

8.2 Sample preparation

No single method of sample preparation can be universally applied due to the diverse nature of samples which are analyzed by 2-D PAGE.

The method is considered to resolve proteins differing by as little as 0.1 pH units in their pI and by 1 kDa in their molecular mass. Therefore, whatever method of sample preparation is employed, it is essential to minimize protein modifications which might result in artifactual spots on the 2-D maps. In particular, samples containing urea must not be heated as this can introduce considerable charge heterogeneity due to carbamylation of the proteins by isocyanate formed in the decomposition of urea. In addition, proteases present within samples can readily produce artifactual spots on 2-D maps, so that samples should be subjected to minimal handling and kept cold at all times. Protease inhibitors such as phenylmethanesulfonyl fluoride (PMSF) can be added, but it must be remembered that such reagents can also modify proteins and introduce charge artifacts. Four broad categories of sample types will be considered: namely, body fluids, solid tissues, circulating and cultured cells, and plant tissues.

8.2.1 Body fluids

Soluble samples available in liquid form can often be analyzed by 2-D PAGE with no or minimal pretreatment. Important examples of this group of proteins are the body fluids such as serum, plasma, urine, cerebrospinal fluid (CSF), semen, prostatic fluid, and amniotic fluid. Serum and plasma samples have a relatively high protein concentration and can be analyzed directly by 2-D PAGE after dilution with sample solubilization buffer (see Section 8.3). A major problem with serum samples, however, is that the very high abundance of proteins such as albumin and immunoglobulins can obscure the minor components on the 2-D maps. This problem can be overcome by depleting these proteins from the sample prior to the 2-D separation using affinity column chromatography [2], but there is always the possibility that this can result in the non-specific removal of other protein components.

Other types of body fluid samples have a low protein concentration and contain relatively high concentrations of salts which can interfere with the IEF dimension. Such samples must usually be desalted, prior to 2-D PAGE, by techniques such as dialysis or gel chromatography. The samples must then be concentrated by methods such as lyophilization, dialysis against polyethylene glycol, precipitation with TCA, or precipitation with ice-cold acetone [2]. The latter method has been found to be particularly satisfactory for the concentration of a variety of samples for 2-D analysis [8]. In addition, the sensitive detection methods currently available, such as silver staining (see Section 9.6), have made it possible to analyze body fluid samples without recourse to extensive desalting and

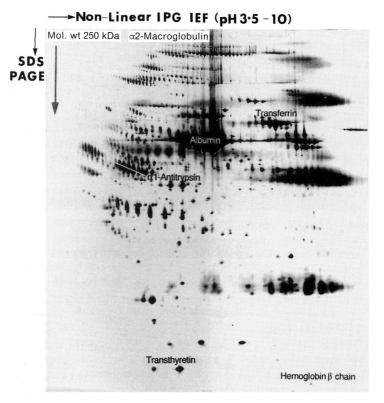

FIGURE 8.1: 2-D PAGE separation of 45 µg of human plasma proteins visualized by silver staining. A non-linear pH 3.5–10 IPG separation was used in the first dimension. The second dimension separation was 9–16%T gradient SDS–PAGE. Reproduced from reference [27] with permission from VCH.

concentration. An example of serum proteins separated by 2-D PAGE is shown in *Figure 8.1*.

8.2.2 Solid tissue samples

Solid tissues are usually disrupted in the presence of solubilization buffer. The recommended procedure is to break up the tissue while it is still frozen, preferably at liquid nitrogen temperature. Small tissue specimens which have been wrapped in aluminum foil and frozen in liquid nitrogen can be crushed to a fine powder between two cooled metal blocks. Large tissue pieces can be processed by homogenization in solubilization buffer using a rotating-blade type homogenizer (e.g. Polytron or Ultra-Turrax), but heating and foaming must be minimized. An example of human heart proteins separated by 2-D PAGE is shown in *Figure 8.2*.

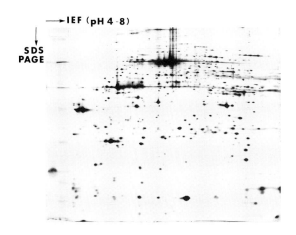

FIGURE 8.2: *2-D PAGE separation of 50 μg human heart proteins visualized by silver staining. The first dimension IEF tube-gel contained pH 4–8 synthetic carrier ampholytes and the second dimension gel was 12%T SDS–PAGE.*

8.2.3 Circulating and cultured cells

Circulating cells (e.g. erythrocytes, leukocytes, platelets) and cells grown *in vitro* in suspension culture can simply be harvested by centrifugation, washed in phosphate-buffered saline (PBS) or balanced salt solution (BSS) and solubilized in sample solubilization buffer. Samples can then be stored at $-70°C$ prior to 2-D analysis. Cells cultured *in vitro* on glass or plastic substrates also require little preparation. The culture medium should be carefully aspirated and the cell layer washed with PBS, BSS or an isotonic sucrose solution to minimize contamination of the sample with medium, particularly serum proteins. The cell layer can then be scraped off, but the use of proteolytic enzymes should be avoided. Alternatively, the cells can be lysed directly while still attached to the substrate by the addition of a small volume of solubilization buffer. If the cells contain high levels of nucleic acids, the samples should be treated with a protease-free DNase/RNase mixture to minimize sample viscosity [9,10]. An example of proteins of human skin fibroblasts separated by 2-D PAGE is shown in *Figure 8.3*.

8.2.4 Plant tissues

The majority of plant proteins, such as seed and membrane preparations, can be treated in a similar way to animal tissues. However, leaf proteins must first be extracted with acetone to remove phenolic pigments [2].

8.3 Sample solubilization

The ideal solubilization procedure for 2-D PAGE would result in the disruption of all non-covalently bound protein complexes and aggregates into a solution of individual polypeptides. If this is not

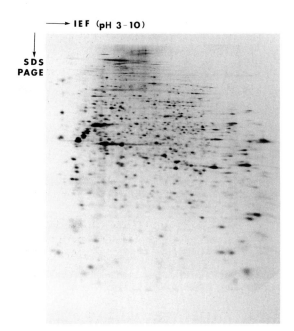

FIGURE 8.3: 2-D PAGE separation of $[^{35}S]$-methionine-labeled human skin fibroblast proteins visualized by autoradiography. The first dimension flat-bed IEF gel contained a mixture of synthetic carrier ampholytes in the pH 3–10 range. The second dimension was 8–20%T non-linear gradient SDS–PAGE.

achieved, persistent protein complexes in the sample are likely to result in new spots in the 2-D pattern, with a concomitant reduction in the intensity of those spots representing the single polypeptides.

The most widely used solubilization procedure is that described by O'Farrell [6], using a mixture of 4% (w/v) NP-40, 9.5 M urea, 1% (w/v) DTT, and 2% (w/v) synthetic carrier ampholytes. This method gives excellent results for the majority of samples. However, not all protein complexes are fully disrupted by this mixture and it is often found that not all sample proteins enter the first dimension IEF gels.

As discussed in Chapter 6, SDS is able to disrupt most non-covalent protein interactions, but its anionic character precludes its inclusion in IEF gels. However, SDS can be used as a pre-solubilization procedure for samples prior to 2-D PAGE. In this approach, the sample is initially solubilized in 1% (w/v) SDS and then diluted with the standard NP-40/urea mixture. The aim here is to displace the SDS from the proteins and replace it with the non-ionic detergent NP-40, thereby maintaining the proteins in a soluble state. The ratio of SDS to protein (1:3) and SDS to NP-40 (1:8) must be carefully controlled for effective solubilization while minimizing the deleterious effects of SDS on IEF.

A variety of alternative procedures for sample solubilization for 2-D PAGE have been described and reviewed in reference [11]. Many of these methods were developed for the analysis of specific types of samples and have not gained acceptance as general procedures [11]. Of more general applicability is the zwitterionic detergent, CHAPS (see Section 5.4.2). CHAPS is particularly effective for the solubilization of membrane proteins [12] and is now often used in preference to NP-40 for the solubilization of proteins for 2-D PAGE. Rabilloud and his colleagues have developed a series of amidosulfobetaine detergents (see Section 5.4.2) which have been found to be useful solubilization agents for samples to be analyzed by 2-D PAGE [13].

8.4 The first dimension

The first dimension of 2-D PAGE separates the proteins according to their charge properties by IEF. This technique is discussed in detail in Chapter 7, and only the features of this technique specific to 2-D PAGE will be discussed here.

8.4.1 IEF using synthetic carrier ampholytes

The first dimension of 2-D PAGE as described by O'Farrell [6] is carried out in cylindrical rod IEF gels (3–5%T) cast in glass capillary tubes (typically 1–1.5 mm i.d.). The gels contain 8 M urea and a non-ionic (e.g. NP-40) or zwitterionic (e.g. CHAPS) detergent. The detergent is usually included at 2% (w/v), but this can often be reduced to 0.5% (w/v) without any adverse effect on the separation pattern [2]. Recently, 2% (w/v) octyl-β-D-glucopyranoside has been recommended, particularly for inclusion in first dimension IEF gels to be used in preparative 2-D PAGE applications (M. Pluskal, personal communication).

First dimension IEF gels also contain 2% (w/v) synthetic carrier ampholytes which are responsible for the generation and maintenance of the pH gradient, on which the quality of the 2-D separation depends. As discussed in Section 7.2.4, a variety of synthetic carrier ampholyte preparations are available, and most of these can be used for 2-D PAGE. However, it must be pointed out that few suppliers submit their ampholyte preparations to quality control by 2-D PAGE, so that batch variability (see Section 7.2.8) can result in problems of reproducibility in 2-D patterns. A broad range (pH 3–10) ampholyte is typically employed, but many investigators now prefer to use a pH 4–8 ampholyte mixture, as the final pH gradients rarely extend beyond pH 7.5 using rod gels (see Section 7.2.4).

Rod gels can be readily prepared by sealing the bottom of the tubes with Parafilm and introducing the gel solution with a long, blunt-ended, narrow-gauge needle or a metal filling cannula. The needle should be initially inserted to the bottom of the tube and then slowly withdrawn to avoid trapping any air bubbles. It is essential that the tubes should be filled to the same level in order to maximize reproducibility. After polymerization, the gels are mounted in a suitably modified vertical gel apparatus. In the standard procedure, the apparatus is arranged with the cathodic electrolyte (typically 20 mM NaOH) in the upper buffer reservoir and the anodic electrolyte (typically 10 mM H_3PO_4) in the lower reservoir. The gels are usually prefocused, typically at 200 V for 15 min, 300 V for 30 min and finally 400 V for 60 min for a total of 600 Vh.

Samples are then applied using a microsyringe to the top of the IEF gels and under a thin layer of buffer containing urea, detergent and ampholytes to protect the proteins from the basic pH of the cathodic electrolyte. Radioactive samples are usually loaded according to total counts. Non-radioactive samples should be loaded according to protein content. For silver staining a load of 10–50 µg is generally appropriate, while 100–1000 µg are required for staining with Coomassie brilliant blue. Most protein assays are subject to interference by the combination of urea and detergent present in the standard solubilization mixture. Samples solubilized in 1% (w/v) SDS can be assayed using the Bradford dye-binding assay [14].

As discussed in Section 7.2.6, the optimal run time for IEF must be determined for each individual sample type to be analyzed; it is also dependent on the length of the separating gel used. Tube IEF gels are normally run at 800–1000 V with typical run times being 18 h (18 000 Vh) for an 18 cm gel and 35 h (35 000 Vh) for a 25 cm gel.

8.4.2 Non-equilibrium pH gradient electrophoresis

As discussed in detail in Section 7.2.8, pH gradients generated using synthetic carrier ampholytes are subject to severe cathodic drift and consequent instability. This problem is exacerbated in the case of rod IEF gels, as a result of the very high electroendosmotic flow caused by charged groups on the glass walls of the capillary tubes. As a result, pH gradients rarely extend above pH 8.0, leading to the loss of basic proteins from 2-D maps. Various approaches to this problem have been described (reviewed in reference [2]) but the most commonly used method is again attributable to O'Farrell [15] and is known as non-equilibrium pH gradient electrophoresis (NEPHGE). In this procedure, the IEF apparatus is assembled with the lower reservoir filled with the basic catholyte solution and the upper reservoir is filled with the acidic anolyte solution. No prefocusing step is used, the samples are applied at the anodic end and much shorter run times in the order of 4000 Vh are used. Few proteins will reach their pI under these conditions, so that separation occurs on the basis of protein mobility in the presence of a rapidly forming pH gradient. Such transient state, non-equilibrium focusing is difficult to control reproducibly and is very sensitive to experimental conditions, carrier ampholytes, run time, gel length, and sample composition [11]. Considerable care must be taken to ensure that samples are analyzed by NEPHGE under identical conditions in order to obtain comparable 2-D maps. A typical 2-D separation using the NEPHGE technique is shown in *Figure 8.4*. Acidic proteins (pI < 7) are not separated by this technique, so that a combination of 2-D patterns obtained using equilibrium IEF and NEPHGE should be used to analyze each sample fully (see *Figure 8.4*).

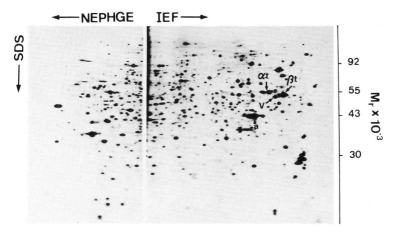

FIGURE 8.4: *2-D PAGE separation of acidic (IEF) and basic (NEPHGE) polypeptides of asynchronous HeLa cells. In IEF the pH range is from pH 7.5 (left) to 4.5 (right). In NEPHGE the pH range is from pH 7.5 (right) to 9.5 (left). The labeled proteins are: a, actin; αt, α-tubulin; βt, β-tubulin; v, vinculin. Reproduced from reference [10] with permission from Academic Press.*

8.4.3 Immobilized pH gradients

In Section 7.3.1 we discussed how the limitations inherent in the use of synthetic carrier ampholytes for the generation of pH gradients can largely be overcome using immobilized pH gradients (IPG). Soon after the introduction of IPGs, several investigators began to adapt this technology to the first dimension separation of 2-D PAGE. Although it is possible to prepare IPG IEF gels in glass capillary tubes, it is better and simpler to use horizontal slab IPG IEF gels cast on plastic supports, as described in Section 7.3.4. However, early attempts to use IPG IEF for 2-D PAGE resulted in various problems, reviewed in reference [2]. These included those problems associated with the use of wide-range IPGs, which can be solved using hybrid or mixed-bed IEF, where low concentrations (e.g. 0.5%, w/v) of synthetic carrier ampholytes are added to the IPG gel (see Section 7.3.2).

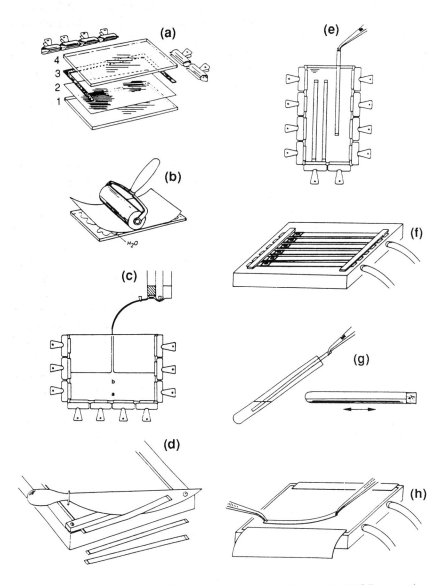

FIGURE 8.5: *Schematic outline of the horizontal IPG 2-D PAGE procedure. (**a**) Polymerization cassette for preparation of IEF and SDS–PAGE gels on plastic backings: (1, 4) glass plates, (2) GelBond PAG film, (3) gasket. (**b**) GelBond PAG film is rolled onto the glass plate with a few drops of water. (**c**) Preparation of IPG gel using a two-chamber gradient mixer. (**d**) The washed and dried IPG gel is cut into strips. (**e**) Rehydration of IPG strips. (**f**) Set-up of first dimension horizontal flat-bed apparatus. (**g**) Equilibration of first dimension IPG strips. (**h**) Second dimension horizontal flat-bed SDS–PAGE. Reproduced from reference [16] with permission from VCH.*

In addition, severe vertical streaking of the 2-D patterns was often observed. This phenomenon was found to be associated with difficulties in elution and transfer of proteins from the first dimension IPG gels to the second dimension SDS–PAGE gels. Görg and her colleagues [16] attributed this phenomenon to the presence of fixed charges on the Immobiline gel matrix leading to increased electroendosmosis in the region of contact between the first and second dimension gels, resulting in disturbance in the migration of proteins between the two dimensions. This problem was overcome by the development of a modified equilibration step between the first and second dimensions (see Section 8.7).

The procedures of 2-D PAGE using IPG IEF gels in the first dimension currently recommended are largely based on the work of Görg and her colleagues [16]. The procedure is outlined diagrammatically in *Figure 8.5*. IPG slab gels of 0.5 mm thickness are cast on GelBond PAG film (Pharmacia–LKB) either in a large format (25 cm wide with an 18 cm pH gradient separation distance) or smaller sizes (e.g. 11 cm or 4 cm pH gradient separation distances) by the cassette technique outlined in Section 7.3.4 (*Figure 8.5a–c*). For most 2-D applications a 3–4 pH unit range IPG is recommended, for example, pH 4–8 for acidic and neutral polypeptides (*Figure 8.6*), and pH 6–10 for neutral and basic polypeptides (*Figure 8.7*). However, it is also possible to use wider (e.g. pH 3–10.5) or narrower (<1 pH unit) pH gradients. The pH gradients used are generally linear, but is also possible to employ non-linear IPGs to optimize 2-D separations of particular types of sample proteins (*Figure 8.8*). After polymerization is complete, the gels are washed, dried and stored at $-20°C$ as described in Section 7.3.4. As an alternative, commercially available ready-made IPG gels (Immobiline DryPlate or DryStrip, Pharmacia) can be used for 2-D PAGE.

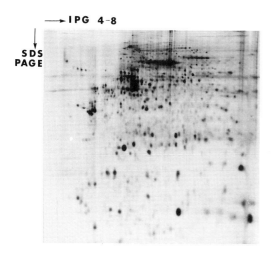

FIGURE 8.6: 2-D PAGE separation of 50 µg human heart proteins using pH 4–8 IPG IEF in the first dimension. The second dimension was a vertical 12%T SDS–PAGE. The pattern was visualized by silver staining.

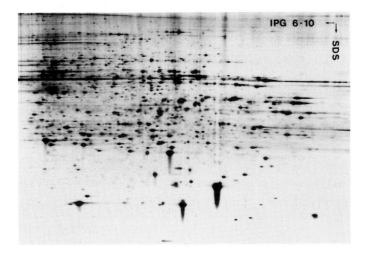

FIGURE 8.7: 2-D PAGE separation of 50 μg human heart proteins using pH 6–10 IPG IEF in the first dimension. The second dimension was a horizontal 12–15%T gradient SDS–PAGE. The pattern was visualized by silver staining. Photograph courtesy of Dr Angelika Görg.

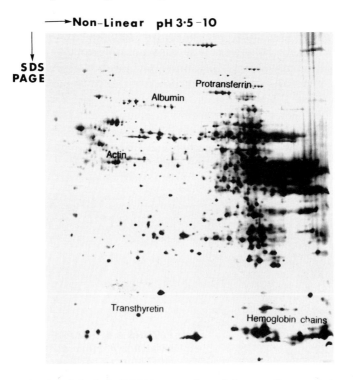

FIGURE 8.8: 2-D PAGE separation of human liver proteins using non-linear pH 3.5–10 IPG IEF in the first dimension. The second dimension was a vertical 9–16%T gradient SDS–PAGE. The pattern was visualized by silver staining. Reproduced from reference [28] with permission from VCH.

Prior to running the first dimension, the desired number of gel strips (5 or 3 mm in width) are cut off the slab with a paper cutter (*Figure 8.5d*) and rehydrated to their original thickness (0.5 mm) with a solution containing 8 M urea, non-ionic or zwitterionic detergent, DTT, and synthetic carrier ampholyte (if required) (*Figure 8.5e*). The rehydrated strips are placed side by side, 2 mm apart, on the cooling platen of the horizontal flat-bed apparatus (*Figure 8.5f*). The electrode strips are soaked with 10 mM glutamic acid (anode) and 10 mM lysine (cathode), or alternatively distilled water can be used for both electrodes.

Samples are best applied into silicon rubber frames placed on the surface of the strips. They are usually applied either at the anodic or cathodic end of the gel, but can, in fact, be applied at any point on the gel surface. The optimal sample application point should, of course, be determined for the particular sample type being analyzed. For improved entry of sample proteins, the applied voltage is limited to 15 V/cm for 30 min and then 30 V/cm for 1 h. IEF is then continued with the maximum settings of 3500 V, 2 mA and 5 W until constant focusing patterns are obtained. Running conditions depend on the sample, pH gradient, whether synthetic carrier ampholytes are present and the separation distance. Some typical values are given in *Table 8.1*.

TABLE 8.1: Running conditions for IPG gels for 2-D PAGE

Temperature:	20°C		
Current max:	0.05 mA per strip		
Power max:	3.0–5.0 W		
Voltage max:	3500 V		
	Time	Voltage	Vh
Sample entry	30 min	150 V	75
	60 min	300 V	300

Total run times (3500 V max)

	11 cm gel		18 cm gel	
pH gradient	Time	Vh	Time	Vh
4–7	7 h	24000	12 h	42000
4–9	5 h	17000	8 h	28000
6–10	6 h	21000	10 h	35000
3–10.5	3 h	10500	5 h	17500

8.5 Gel recovery

After the first dimension separation is complete, cylindrical IEF gels can be removed from the glass tubes by injecting water around the gels or by application of water or air pressure (see Section 9.1). Care must be taken not to damage, stretch or break the rather delicate gels. To overcome this problem, in the commercial 2-D PAGE system of Millipore [17], the IEF gel is polymerized around a 0.08 mm thread which is incorporated into the length of the glass tubes. In addition, a modified Luer adapter is provided to connect the IEF tube to a syringe for gel extrusion. Recovery of IPG gel strips is much simpler as they can simply be lifted from the flat-bed apparatus while still on their plastic supports.

If the IEF gels are to be used immediately for the second dimension, they can be extruded directly into equilibration buffer (see Section 8.7). If not used immediately, the gels should be wrapped in Parafilm and stored frozen at $-80°C$ until required.

8.6 Estimation of pH gradients

Protein pI values can be used to characterize spot positions on 2-D maps, but this requires that the pH gradient formed during IEF must be calibrated with accuracy. Using cylindrical IEF gels, the pH gradient can be determined by transversely slicing the gel into thin segments, eluting them into a small volume of 10 mM KCl, and measuring the pH of the resulting solution. However, this approach is complicated by the effects of urea and temperature on pH (see Section 7.2.7). This method of direct pH measurement cannot be used with IPG gels (see Section 7.3.7). The best approach is to use a set of marker proteins of established pI. However, most commercial pI marker kits are not suitable for 2-D applications as they often contain multimeric proteins which would be dissociated under the denaturing conditions used in 2-D PAGE. Undoubtedly the best method currently available for calibrating pH gradients in first dimension IEF gels is the use of carbamylated charge standards as described in Section 7.2.7. An example of the use of these standards is shown in *Figure 8.9*.

8.7 Equilibration between dimensions

An equilibration step is usually used between the two dimensions of 2-D PAGE. Cylindrical tube IEF gels are incubated for a few minutes at room temperature in 0.125 M Tris buffer, pH 6.8, containing 2% (w/v) SDS under reducing conditions. More vigorous equilibration conditions are used for IPG IEF gels which are incubated for 15 min in 50 mM Tris buffer, pH 8.8, containing 2% (w/v) SDS, 1% (w/v) DTT, 6 M urea, and 30% (w/v) glycerol, and then for 15 min in the same solution containing 5% (w/v) iodoacetamide in place of DTT [16]. Iodoacetamide is added to

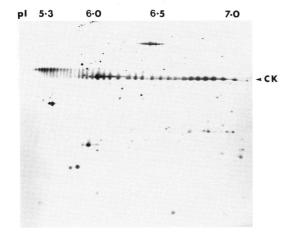

FIGURE 8.9: 2-D PAGE separation showing carbamylated creatine kinase pI markers visualized by silver staining. The sample also contained a small amount of human heart proteins to give reference points for the comparison and matching of other 2-D patterns. The relative pI values indicated were estimated from the known pI values of the charge train proteins.

the second portion of the equilibration buffer to eliminate non-sample related staining artifacts, such as the short horizontal and vertical streaks ('point streaking') which can occur during silver staining [16].

The function of the equilibration step is to allow the proteins within the IEF gel matrix to interact fully with SDS so that they migrate with the proper characteristics in the second dimension. Diffusion during equilibration results in broadening of protein zones and loss of some protein from the gel. This has tempted some investigators to dispense with this step, but the result can be streaking, especially of the high molecular mass proteins. Moreover, equilibration of IPG IEF gels is essential to ensure good transfer of the proteins between dimensions and to minimize vertical streaking of the 2-D patterns.

8.8 Transfer between dimensions

First dimension IEF gels must be applied, after equilibration, to the origin of the second dimension SDS–PAGE gels. In order to obtain good 2-D separations, it is essential that a good contact is made between the two gels over their entire lengths.

In most procedures using cylindrical first dimension IEF gels, the gels are cemented in place on the top of the vertical SDS–PAGE slab gels using buffered agarose. However, many commercial agarose preparations are contaminated with impurities which result in artifacts on the 2-D patterns when silver staining is used. It is usually possible, provided that the thickness of the SDS–PAGE gels is matched to the that of the IEF gels, to apply the IEF gels directly to the second dimension slab gels without the use of agarose, as the IEF gels become bonded during electrophoresis to the surface of the SDS gel. Care must be taken when handling cylindrical gels to avoid any stretching which will result in distortions of the 2-D patterns.

IPG strips are much easier to handle as they are bound to a flexible plastic support. These strips can simply be applied to the top of vertical SDS–PAGE slab gels and cementing with agarose is usually unnecessary. Using horizontal flat-bed second dimension SDS–PAGE gels, the IPG strips are transferred, gel-side down, on to the surface of the SDS gel parallel to the cathodic electrode wick and separated from it by 1 mm (*Figure 8.5h*).

8.9 The second dimension

SDS–PAGE using the discontinuous buffer system of Laemmli [7] is used almost universally for the second dimension of 2-D PAGE. Gels of either a single polyacrylamide concentration or containing a linear or non-linear polyacrylamide concentration gradient, to extend the range over which proteins of different molecular mass can be effectively separated, may be used (see Chapters 3 and 6). Stacking gels are often used, but these can generally be omitted as the protein zones within IEF gels are already concentrated and the non-restrictive IEF gel tends to act as a stacking gel. Gels are prepared using standard procedures (see Section 3.7.2), but it is advantageous to cast gels in batches, rather than individually, as this improves reproducibility of 2-D patterns. Vertical SDS–PAGE gel systems are usually used for 2-D PAGE, but it is possible to use horizontal SDS–PAGE gels, particularly when IPG IEF gels are used in the first dimension [16]. In the latter case, the SDS–PAGE gels are cast on plastic supports (GelBond PAG) [16].

8.10 Molecular mass standards

In the same way that pI markers are essential for the calibration of the IEF dimension (see Section 8.6), so a series of M_r standards is required to provide a reference for the SDS–PAGE dimension. These standards are usually a mixture of purified proteins of known M_r which are electrophoresed down one side of the SDS–PAGE gels. Several kits of such markers, both radiolabeled and non-radioactive, are available commercially.

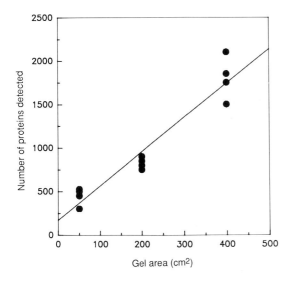

FIGURE 8.10: *Plot to show the influence of gel area on the number of proteins which can be resolved by 2-D PAGE. Redrawn from reference [17] with permission from Eaton Publishing.*

8.11 Gel size, resolution and reproducibility

The resolving capacity of 2-D gels is dependent on the separation length in both dimensions and is, therefore, dependent on the area of the 2-D gels. Using standard 2-D gel formats, measuring 16–20 cm in each dimension, between 1000 and 2000 proteins can be resolved, whereas only a few hundred protein spots can be resolved using minigel formats. This dependence of resolution on gel area (*Figure 8.10*) has prompted some investigators to adopt very large formats, for example, 3000–4000 polypeptides can be resolved using 32×43 cm 2-D gels [18]. An example of such a separation is shown in *Figure 8.11*.

The ability to generate reproducible 2-D patterns is as important as the absolute protein resolving capacity. It is, therefore, essential to perform all the steps involved in 2-D PAGE in as careful and controlled a manner as possible to maximize reproducibility. Reproducibility is also improved if as many gels as possible are prepared and electrophoresed simultaneously. This has resulted in the development of two dedicated 2-D PAGE systems which are available commercially.

The first system is based on the pioneering work of Norman and Leigh Anderson and their co-workers, first at the Argonne National Laboratory and later at the Large Scale Biology Corporation [19,20]. This apparatus is available commercially through Hoefer and is described in detail in reference [21]. Two gel formats are supported (18×18 cm and 25×20 cm) and batches of 20 gels can be processed at

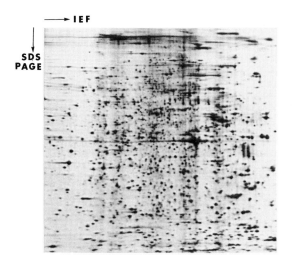

FIGURE 8.11: Large format (37 × 39 cm) 2-D PAGE separation of [^{35}S]methionine-labeled mouse embryonic fibroblast proteins, showing the large number (approximately 3000) of proteins which can be resolved using this technique. The pattern was visualized by autoradiography. Reproduced from reference [29] with permission from the American Association for Clinical Chemistry.

the same time. Some of the equipment is shown in *Figure 8.12*. A notable feature of this apparatus is that the second dimension SDS–PAGE gels are rotated through 90° and are electrophoresed from side to side.

The second system, providing a 22 × 22 cm format, is available from Millipore and is described in detail in reference [17]. Either 15 analytical (1 mm i.d.) or eight preparative (3 mm i.d.) cylindrical IEF gels can be run in the first dimension and five SDS–PAGE slab gels (1 mm thick) can be electrophoresed in the second dimension. Notable features of this equipment include the provision of a programmable multiple-output power supply, which allows the simultaneous control of both the IEF and SDS–PAGE dimensions, and a Peltier cooling device which facilitates accurate control of temperature in the SDS–PAGE dimension. Some components of the Millipore 'Investigator' system are shown in *Figure 8.13*.

8.12 Two-dimensional PAGE under native conditions

2-D electrophoresis under native conditions is limited in its application to soluble protein samples, but it can be used to advantage for the investigation of native physico-chemical properties and biological activities of proteins. The technique which is most commonly used is based on the 2-D combination of native IEF in the first dimension with PAGE in the absence of urea and detergents in the second dimension. This methodology was developed by Dale and Latner [22] and has been extensively exploited by Manabe [23,24] and others [25] for the analysis of body fluids, such as serum (*Figure 8.14*), cerebrospinal fluid, and saliva.

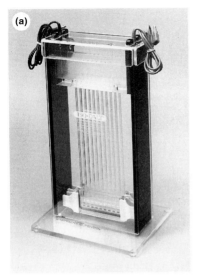
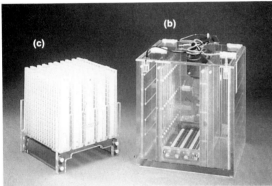

FIGURE 8.12: Some of the components of the Iso-Dalt 2-D PAGE system. (**a**) ISO tube IEF gel unit, (**b**) 10-place DALT multiple SDS–PAGE gel unit, (**c**) Western blotting unit for the DALT tank. Photographs courtesy of Hoefer Scientific Instruments.

8.13 Two-dimensional 2-D PAGE under denaturing conditions

Although the method of O'Farrell [6] is the method of choice for the 2-D separation of the majority of complex protein mixtures, certain types of sample are not amenable to analysis by this technique. This has resulted in the development of other 2-D methods using denaturing conditions in both dimensions.

8.13.1 Ribosomal proteins

Both prokaryotic and eukaryotic ribosomal proteins are strongly basic and are not amenable to analysis using IEF. This has necessitated the development of special 2-D systems capable of separating these proteins. This problem has been solved using combinations of gels containing urea under acid or basic conditions in the first dimension with SDS–PAGE in the second dimension [2]. Madjar and colleagues have developed a procedure known as the 'method of four corners' which uses a combination of four different 2-D gel systems for the complete analysis of ribosomal proteins [26]. The four systems used are (i) acid–urea vs. SDS–PAGE, (ii) basic–urea vs. SDS–PAGE, (iii) basic–urea vs. acidic–urea, and (iv) acidic–urea vs. acidic–urea. An example of this method applied to the analysis of the ribosomal proteins of Chinese hamster ovary cells is shown in *Figure 8.15*.

8.13.2 Nuclear proteins

The basic character of histone proteins, the major protein components of chromosomes, and the similarity of their molecular masses makes them difficult to separate by the 2-D PAGE method of O'Farrell [6]. Several methods based on combinations of acid–urea gels with SDS–PAGE have

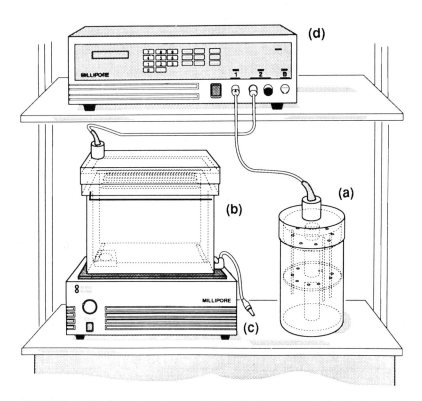

FIGURE 8.13: The Investigator 2-D PAGE system. *(a)* Tube IEF gel unit, *(b)* SDS–PAGE unit, *(c)* Peltier cooling system, *(d)* computer-controlled power pack. Diagram courtesy of Millipore.

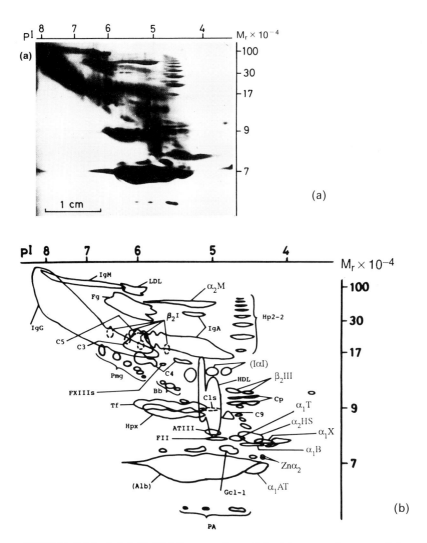

FIGURE 8.14: (a) 2-D PAGE separation of 2 μl human plasma proteins using native conditions followed by Coomassie brilliant blue staining. (b) Reference map of human plasma proteins separated by native 2-D PAGE. Reproduced from reference [24] with permission from VCH.

been described for this group of proteins [2]. The best method which has been described to date uses a combination of 15%T polyacrylamide gels containing 5% acetic acid and 2.5 M urea in the first dimension, with second dimension gels containing 6.5 M urea and 6 mM Triton X-100 [8]. A typical separation obtained with this technique is illustrated in *Figure 8.16*.

Non-histone proteins can generally be analyzed by the 2-D PAGE method of O'Farrell [6], but nucleolar proteins are best separated using a 2-D combination of acetic acid–urea and SDS–PAGE gels [2].

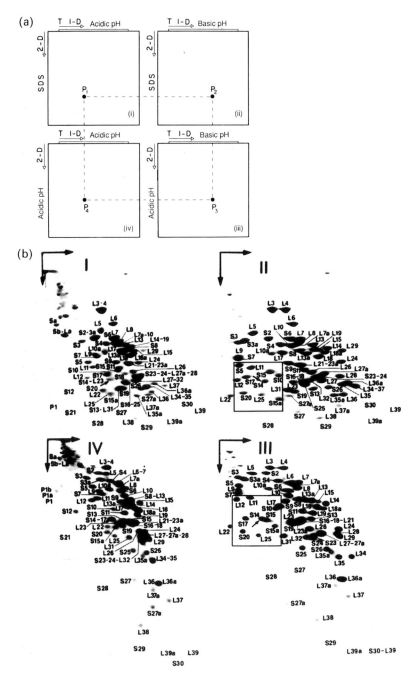

FIGURE 8.15: (**a**) Schematic representation of the position of one given protein in the four different 2-D gel systems used for the method of four corners separation of ribosomal proteins. Reproduced from reference [26] with permission from Academic Press. (**b**) 2-D separation of Chinese hamster ovary cell ribosomal proteins using the method of four corners. Reproduced from reference [30] with permission from the American Society for Microbiology.

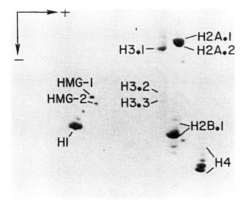

FIGURE 8.16: 2-D separation of calf thymus histone proteins, The first dimension was 15%T SDS–PAGE. The second dimension gel contained 0.9 M acetic acid, 6 M urea, 0.375% (w/v) Triton X-100. Reproduced from reference [31] with permission from Academic Press.

8.13.3 Membrane proteins

Membrane proteins can be analyzed by the O'Farrell 2-D PAGE technique [6], but the hydrophobic character of many of these proteins can cause problems in solubility during the IEF dimension and result in smearing, ill-defined spots and loss of proteins from the 2-D patterns. This problem has been approached using polyacrylamide gels containing alternative solubilizing agents such as Triton X-100, Triton C10 or chloral hydrate in the first dimension in combination with SDS–PAGE in the second dimension [2]. However, these methods do not generally result in high resolution separations as size differences in the proteins contribute to the separation in both dimensions, resulting in a diagonal distribution of protein spots.

References

1. Smithies, O. and Poulik, M.D. (1956) *Nature*, **177**, 1033–1035.
2. Dunn, M.J. (1987) in *Advances in Electrophoresis* (A. Chrambach, M.J. Dunn and B.J. Radola, eds). VCH, Weinheim, Vol. 1, pp. 1–109.
3. Iborra, G. and Buhler, J.M. (1976) *Anal. Biochem.*, **74**, 503–511.
4. Klose, J. (1975) *Humangenetik*, **26**, 231–243.
5. Scheele, G.A. (1975) *J. Biol. Chem.*, **250**, 5375–5385.
6. O'Farrell, P.H. (1975) *J. Biol. Chem.*, **250**, 4007–4021.
7. Laemmli, U.K. (1970) *Nature*, **227**, 680–685.
8. Pipkin, J., Anson, J.F., Hinson, W.G., Burns, E.R. and Wolff, G.L. (1985) *Electrophoresis*, **6**, 306–313.
9. Garrels, J.I. (1979) *J. Biol. Chem.*, **254**, 7961–7977.
10. Bravo, R. (1984) in *Two-Dimensional Gel Electrophoresis of Proteins* (R. Bravo, and J.E. Celis, eds). Academic Press, Orlando, FL, pp. 3–36.
11. Dunn. M.J. and Burghes, A.H.M. (1983) *Electrophoresis*, **4**, 97–116.
12. Perdew, G.H., Schaup, H.W. and Selivonchick, D.P. (1983) *Anal. Biochem.*, **135**, 453–455.

13. Rabilloud, T., Gianazza, E., Cattò, N. and Righetti, P.G. (1990) *Anal. Biochem.*, **185**, 94–102.
14. Bradford, M.B. (1976) *Anal. Biochem.*, **72**, 248–254.
15. O'Farrell, P.Z., Goodman, M.M. and O'Farrell, P.H. (1977) *Methods Cell Biol.*, **12**, 1133–1142.
16. Görg, A., Postel, W. and Günther, S. (1988) *Electrophoresis*, **9**, 531–546.
17. Patton, W.F., Pluskal, M.G., Skea, W.M., Buecker, J.L., Lopez, M.F., Zimmermann, R., Belanger, L.M. and Hatch, P.D. (1990) *BioTechniques*, **8**, 518–527.
18. Levenson, R.M. and Young, D.A. (1991) in *2-D PAGE '91, Proceedings of the International Meeting on Two-Dimensional Electrophoresis* (M.J. Dunn, ed.). National Heart and Lung Institute, London, pp. 12–16.
19. Anderson, N.G. and Anderson, N.L. (1978) *Anal. Biochem.*, **85**, 331–340.
20. Anderson, N.G. and Anderson, N.L. (1978) *Anal. Biochem.*, **85**, 341–354.
21. Anderson, L. (1988) *Two-Dimensional Electrophoresis*. Large Scale Biology Press, Washington, DC.
22. Dale, G. and Latner, A.L. (1969) *Clin. Chim. Acta*, **24**, 61–68.
23. Manabe, T., Hayama, E. and Okuyama, T. (1982) *Clin. Chem.* **28**, 824–827.
24. Manabe, T., Takahashi, Y., Higuchi, N. and Okuyama, T. (1985) *Electrophoresis*, **6**, 462–467.
25. Marshall, T. and Williams, K.M. (1991) *Electrophoresis*, **12**, 461–471.
26. Madjar, J.-J., Michel, S., Cozzone, A.J. and Reboud, J.-P. (1979) *Anal. Biochem.*, **92**, 174–182.
27. Hughes, G., Frutiger, S., Paquet, N., Ravier, C., Pasquali, C., Sanchez, J.-C., James, R., Tissot, J.-D., Bjellqvist, B. and Hochstrasser, D.F. (1992) *Electrophoresis*, **13**, 707–714.
28. Hochstrasser, D.F., Frutiger, S., Paquet, N., *et al.* (1992) *Electrophoresis*, **13**, 992–1001.
29. Young, D.A. (1984) *Clin. Chem.*, **30**, 2104–2108.
30. Madjar, J.-J., Frahm, M., McGill, S. and Roufa, D. (1983) *Mol. Cell. Biol.*, **3**, 190–197.
31. Davie, J.R. (1982) *Anal. Biochem.*, **120**, 276–281.

9 Detection Methods

9.1 Gel recovery

After electrophoresis is complete, the gel is removed from the apparatus for localization of the separated polypeptides. Gloves should be worn during procedures involving any handling of the gels, as contact with the skin can lead to the production of artifacts during staining, particularly in the case of sensitive methods such as silver staining.

Slab gels are simply dismantled by removing the spacers and carefully prizing the glass plates apart with a spatula. Gels which have been cast on plastic supports present no problems during handling. Unsupported gels must be handled with care, particularly if they are of a relatively low (< 10%T) acrylamide concentration.

A more severe problem is presented by rod gels which have been cast in glass tubes. Gels cast in capillary tubes, such as those used for the first IEF dimension of 2-D PAGE (see Sections 8.4 and 8.5), are best removed by hydrostatic pressure, which is achieved by attaching a syringe filled with water to the glass tube with silicone rubber tubing (*Figure 9.1*). If tubes of a larger i.d. are used it is possible to remove the gel by rimming it with water injected from a syringe fitted with a long, fine, blunt needle. As a last resort it is possible to break the glass tube carefully, using a hammer, to release the gel.

9.2 Fixation

Procedures have been described for the direct visualization of unfixed proteins within gels, but these are usually applied only if it is wished to recover the separated components (see Section 12.2). For most visualization techniques it is essential to precipitate and immobilize the separated proteins within the gel and to remove any non-protein components which might interfere with subsequent staining. Gels which are to be used for visualization of the enzymic activities of separated proteins must not be subjected to fixation (see Section 9.11).

The best general purpose fixative is 20% (w/v) trichloroacetic acid (TCA), whilst sulfosalicylic acid (1–20%, w/v) or mixtures of TCA and sulfosalicylic acid (10%, w/v of each) are less efficient. Methanolic solutions of acetic acid (e.g. methanol, distilled water, acetic acid, 9:9:2, v/v/v) are very popular for gel fixation, but it should be remembered that low molecular weight species, basic proteins and glycoproteins may not be adequately fixed by this procedure. Aqueous solutions of reagents such as 5% (w/v) formaldehyde or 2% (w/v) glutaraldehyde can be used to cross-link proteins covalently to the gel matrix, but this is not a commonly used approach.

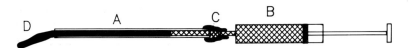

FIGURE 9.1: Extrusion of polyacrylamide gel from a glass tube. A, Glass tube containing the gel; B, water filled syringe; C, silicon rubber tubing connector; D, partially extruded gel.

9.3 General protein stains

In the early days of electrophoresis, methods for the detection of separated protein zones were limited and insensitive, for example: (i) studies of colored proteins, (ii) u.v. absorption, and (iii) Schlieren optical methods. The subsequent development of organic dyes able to react with proteins soon made stains such as Bromophenol blue and Amido black 10B popular. However, it is the Coomassie brilliant blue dyes, originally developed as acid wool dyes, which have been adopted as the most popular general protein staining reagents due to their relatively high sensitivity of detection (approximately 0.2–0.5 µg of protein per band). The level of sensitivity has been further increased with the development of autoradiographic and fluorographic methods for radiolabeled proteins, and the introduction of fluorescent (approximately 10 ng of protein per band) and silver (approximately 0.1 ng of protein per band) staining techniques. A good general review of protein detection methods can be found in reference [1].

9.4 Coomassie brilliant blue

The most popular general protein staining procedures following electrophoresis are based on the use of non-polar, sulfated triphenylmethane Coomassie stains, developed for the textile industry. Coomassie brilliant blue (CBB) R-250 is most often used and requires an acidic medium for electrostatic interaction between the dye molecules and the amino groups of proteins. Staining is usually carried out using 0.1% (w/v) CBB R-250 in methanol, distilled water, and acetic acid (9:9:2, v/v/v). The time required for staining depends on the thickness of the gel and its polyacrylamide concentration. A 10%T gel of 0.5 mm thickness should be stained for about 2 h, whilst the staining time should be increased for thicker and/or more concentrated (i.e. higher %T) gels. In practice, immersion of the gel in staining solution overnight is often convenient. After staining, both the gel matrix and the protein bands contained within it are dark blue. In order to visualize the protein bands it is necessary to destain the gel by gentle agitation in the same acid–methanol solution but in the absence of added dye. The time for destaining is also dependent on gel thickness and polyacrylamide concentration. The process can take up to 24 h but can be accelerated by using several changes of the destaining solution. After destaining, protein zones are visualized as intense blue bands on a colorless background (*Figure 9.2*). Stained gels can be stored in 7% (v/v) acetic acid.

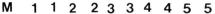

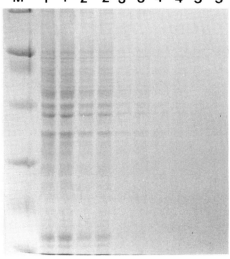

FIGURE 9.2: Separation of human endothelial cell proteins by 10%T SDS–PAGE and visualized using 0.1% (w/v) CBB R-250 in methanol, distilled water, and acetic acid (9:9:2, v/v/v). M, molecular weight markers; 1–5, endothelial cell proteins; 1, 40 µg; 2, 20 µg; 3, 10 µg; 4, 5 µg; 5, 1 µg.

The dimethylated form of the dye, CBB G-250, can be used as a colloidal dispersion in TCA (0.25% (w/v) CBB G-250 in a 50% (v/v) aqueous methanol solution containing 12.5% (w/v) TCA). The advantage of this method is that the colloidal dye particles cannot penetrate the gel matrix and thus selectively form dye–protein complexes. This results in rapid staining of the proteins without the development of background staining. Destaining is, therefore, not required in this procedure and staining can be accelerated at elevated temperatures, taking only a few minutes at 60°C.

Staining methods using CBB are simple and convenient, but suffer from the disadvantage that they are rather insensitive, being able to detect down to 0.2–0.5 μg of protein per band. Thus, for satisfactory staining of a complex protein mixture containing in the order of 100–200 components, it is necessary to load 50–100 μg of total sample protein to the gel. In an attempt to overcome this problem, Neuhoff and co-workers have recently tested 600 modifications of the CBB staining procedure [2] and developed an optimal staining technique using 0.1% (w/v) CBB G-250 in 2% (w/v) phosphoric acid, 10% (w/v) ammonium sulfate, and 20% (v/v) methanol [3]. Staining should be carried out in sealed dishes to minimize evaporation of methanol. This method utilizes CBB in a colloidal form so that no destaining step is required. After staining to the required sensitivity (8–10 h for 1 mm thick gels), gels should be briefly washed with 25% (w/v) methanol and can be stored in 20–25% (w/v) ammonium sulfate in water. This procedure has a significantly increased sensitivity (0.5–1.0 ng of protein per band) compared with other CBB staining techniques (*Figure 9.3*).

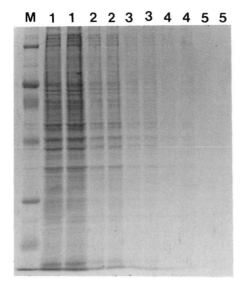

FIGURE 9.3: Separation of human endothelial cell proteins by 10%T SDS–PAGE and visualized using 0.1% (w/v) CBB G-250 in 2% (w/v) phosphoric acid, 10% (w/v) ammonium sulfate and 20% (v/v) methanol. M, molecular weight markers; 1–5, endothelial cell proteins; 1, 40 μg; 2, 20 μg; 3, 10 μg; 4, 5 μg; 5, 1 μg.

9.5 Fluorescent staining methods

An increased sensitivity of detection of proteins in gels can be achieved using fluorescent compounds. Two approaches are possible using these reagents.

In the first approach, proteins are coupled with a fluorescent dye before electrophoresis and fluorescent protein bands are detected after electrophoresis by scanning. Examples of such pre-electrophoretic stains are dansyl chloride and fluorescamine (4-phenylspiro-[furan-2(3H),1-phthalan]-3,3'-dione) which have sensitivities as low as 10 ng and 5 ng of protein per band, respectively. Of particular interest is the compound 2-methoxy-2,4-diphenyl-3(2H)-furanone (MDPF) which can detect as little as 1 ng of protein per band [4].

The main disadvantage of pre-electrophoretic staining procedures is that they can cause protein charge modifications, for example, fluorescamine converts an amino group to a carboxyl group when it reacts with proteins. These modifications will result in altered protein mobility during electrophoresis in buffers which do not contain SDS, resulting in anomalous separations by PAGE, IEF, and 2-D PAGE. To overcome this problem in 2-D PAGE, Jackson et al. [5] developed a procedure for labeling proteins with MDPF while present in the first dimension gel (after IEF), prior to the second dimension separation by SDS–PAGE.

The second approach, which also overcomes the problem of protein charge modifications, is to label the proteins with fluorescent molecules such as 1-aniline-8-naphthalene sulfonate (ANS) and o-phthalaldehyde (OPA) after the electrophoretic separation has been completed. However, these stains are not particularly sensitive (OPA can detect 0.5 µg of protein per band) and the requirement for u.v. illumination or special scanning equipment for their visualization has resulted in a general lack of popularity of this group of procedures.

9.6 Silver staining

The ability of silver to develop images was discovered in the mid-17th century and this property was exploited in the development of photography, followed closely by its use in histological procedures. Silver staining of proteins following electrophoresis was introduced by Switzer et al. in 1979 [6]. Subsequently, over 100 publications have appeared describing variations on the methodology of silver staining, and these are reviewed in references [1] and [7]. The sensitivity of these methods is claimed to be 20–200 times more sensitive than methods using CBB R-250, and they can detect about 0.1 ng of protein per band (*Figure 9.4*).

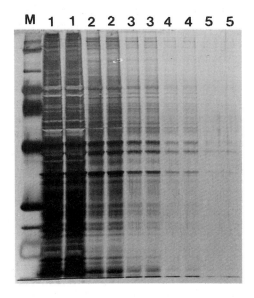

FIGURE 9.4: Separation of human endothelial cell proteins by 10%T SDS–PAGE and visualized by silver staining. M, molecular weight markers; 1–5, endothelial cell proteins; 1, 40 μg; 2, 20 μg; 3, 10 μg; 4, 5 μg; 5, 1 μg.

The high sensitivity of silver staining methods makes them ideal for the detection of trace components within a protein sample or for the analysis of protein samples available in only limited quantity. However, these methods can suffer from a number of disadvantages, for example: (i) high background staining (water and reagent purity is of great importance here), (ii) surface 'mirror' staining, (iii) poor reproducibility, (iv) high cost, (v) slow, labor-intensive procedures and (vi) certain proteins stain poorly, negatively, or not at all. It is recommended, therefore, that silver staining should only be used as a routine staining method when it is essential to take advantage of its high sensitivity. A variety of commercial kits for silver staining are now available from several manufacturers. In the author's experience, these kits vary greatly in their relative sensitivities for protein detection, but their use can certainly alleviate some of the problems of excessive background staining and poor reproducibility which are experienced by many workers using silver staining recipes from the literature.

9.6.1 Fixation

In most published protocols, mixtures of alcohol, acetic acid, and water are recommended for gel fixation. However, TCA is a better general fixative (see Section 9.2) and its use is compatible with silver staining, provided that gels are washed well after fixation to remove the acid. In addition to protein fixation, this step also acts to remove reagents such as Tris, glycine, and detergents (especially SDS) which can bind silver and thereby increase background staining. In this regard, fixation with glutaraldehyde is not recommended as it can result in excessive yellow background due to reaction with glycine from the electrophoresis buffer.

9.6.2 Enhancement

Most procedures include a step prior to silver impregnation of the gel which is designed to increase the sensitivity and/or contrast of the stain. A large variety of pretreatments have been described, but Rabilloud [7] has categorized them into three main approaches. The first method of enhancement (amplification) is by increasing the binding of silver to proteins. A notable example here is the enhanced staining of proteins which have been previously reacted with CBB dyes. The second approach acts to increase contrast by minimizing background staining and is achieved by pretreatment of gels with oxidizing agents such as dichromate and permanganate. This approach can only be used in methods where the gel is acidic up to the development stage (see Section 9.6.3) so that the effect of the oxidizer will be maximal. The third, and most effective approach, is to increase sensitivity by increasing the speed of silver reduction on the proteins. This is achieved by pretreatment of the gel with either sulfhydryl reagents (e.g. thiourea, dithiothreitol, thiosulfate, tetrathionates) or with reducing agents (e.g. glutaraldehyde, formaldehyde, borohydride, dithionite).

9.6.3 Staining and development

All silver staining procedures depend on the reduction of silver and they can be grouped into two types of method depending on the chemical state of the silver ion when used for impregnating the gel.

The first group are alkaline methods based on the use of an ammoniacal silver or diamine solution, prepared by adding silver nitrate to a sodium–ammonium hydroxide mixture. Commonly used procedures of this type are given in references [8–10]. Copper is included in several diamine procedures as it increases its sensitivity and the mechanism of this may be similar to that of the Biuret reaction. The silver ions in the gel complexed to proteins are subsequently developed by reduction to metallic silver with formaldehyde in an acidified environment, usually using citric acid. A typical procedure is detailed in *Table 9.1*.

TABLE 9.1: *Typical silver staining procedure using silver diamine solution [34]*

Step	Procedure	Conditions
1	Fixation	30 min (or overnight) in a solution containing 40% (v/v) ethanol, 10% (v/v) acetic acid
2	Rinsing	15 min in water, followed by 15 min in cold (< 10°C) water
3	Sensitization	30 min in 10% (w/v) glutaraldehyde
4	Rinsing	4 × 15 min in cold water
5	Silver staining	30 min in a solution containing 8 g/l silver nitrate, 40 ml/l 5 N ammonium hydroxide, 40 ml/l sodium hydroxide
6	Rinsing	3 × 5 min in water
7	Development	3–10 min in a solution containing 1 ml/l of 37% (w/v) formaldehyde, 0.1 g/l citric acid
8	Termination	5% (v/v) acetic acid, 20% (v/v) ethanol
9	Storage	Water

In the second group of methods, silver nitrate in a weakly acidic (approximately pH 6) solution is used for gel impregnation. Development is subsequently achieved by the selective reduction of ionic silver to metallic silver by formaldehyde made alkaline with either sodium carbonate or sodium hydroxide. Care must be taken to wash out free silver nitrate from the gel before development to prevent precipitation of silver oxide which may lead to a high background. Recommended procedures for rapid staining (less than 5 h) are given in references [11–13] (for a typical procedure, see *Table 9.2*), but for optimal sensitivity longer methods (see *Table 9.3*) should be used [14, 15].

Silver stains are normally monochromatic, producing a dark brown image. However, if image development is allowed to proceed further, dense protein zones become saturated and color effects can be produced. Some staining protocols have been developed to exploit these color effects, which were claimed to be related to the nature of the polypeptide [16]. However, it is now known that the colors produced depend on: (i) the size of the silver particles, (ii) the refractive index of the gel, and (iii) the distribution of the silver particles within the gel [17]. The colors produced can aid in identification of certain

TABLE 9.2: *Typical procedure for rapid silver staining with silver nitrate [13]*

Step	Procedure	Conditions
1	Fixation	3 × 30 min in a solution containing 40% (v/v) ethanol, 10% (v/v) acetic acid
2	Rinsing	10 min in 20% (v/v) ethanol, followed by 10 min in water
3	Sensitization	1 min in a solution containing 0.3 g/l sodium thiosulfate pentahyrate
4	Rinsing	2 × 1 min in water
5	Silver staining	20–60 min in a solution containing 2 g/l silver nitrate, 250 µl/l of 37% (w/v) formaldehyde
6	Rinsing	10–20 sec in water
7	Development	5–10 min in a solution containing 30 g/l potassium carbonate, 250 µl/l of 37% (w/v) formaldehyde, 10 mg/l sodium thiosulfate pentahydrate
8	Termination	30 min in a solution containing 50 g/l Tris, 20 ml/l acetic acid
9	Storage	Water

TABLE 9.3: *Typical long silver staining method using silver nitrate to obtain high sensitivity* [14]

Step	Procedure	Conditions
1	Fixation	3 × 45 min in a solution containing 40% (v/v) ethanol, 10% (v/v) acetic acid
2	Rinsing	3 × 30 min in water
3	Sensitization	Overnight in a solution containing 0.5% (w/v) glutaraldehyde, 30% (v/v) ethanol, 2 g/l sodium thiosulfate pentahydrate, 68 g/l sodium acetate
4	Rinsing	4 × 30 min in water
5	Silver staining	30–120 min in a solution containing 1 g/l silver nitrate, 250 µl/l of 37% (w/v) formaldehyde
6	Rinsing	10–20 sec in water
7	Development	5–15 min in a solution containing 25 g/l sodium carbonate, 100 µl/l of 37% (w/v) formaldehyde
8	Termination	5–10 min in a solution containing 14.6 g/l EDTA
9	Storage	Water

proteins, but variations in protein concentration can produce color shifts, so confusing identification. A serious disadvantage of these color methods is that saturation and negative staining effects often occur and these result in considerable problems if quantitative analysis is attempted.

All methods of silver staining involve the reduction of ionic silver to its metallic form, but the precise mechanism involved in silver staining of proteins has not been fully established. It has been proposed that silver cations complex with protein amino groups, particularly the ε-amino group of lysine [18], and with sulfur residues of cysteine and methionine [19]. However, Gersten and colleagues [20] have shown that 'stainability' cannot be attributed entirely to specific amino acids and have suggested that some element of protein structure, higher than amino acid composition, is responsible for differential silver staining.

9.7 Radioactive detection methods

For more information of the use, handling and detection of radio-isotopes see reference [21].

9.7.1 Radiolabeling methods

Protein samples to be analyzed by electrophoresis can be radiolabeled synthetically by the incorporation of radioactive amino acids (usually [^{14}C]leucine and/or [^{35}S]methionine). This approach is generally used in tissue culture systems where the culture medium should be depleted of the amino acid used for labeling. It is also possible to radiolabel synthetically the proteins of small pieces of fresh tissue in this way.

Techniques are also available for synthetic labeling of specific proteins, for example, $^{32}PO_4^{3-}$ for phosphoproteins and [^{3}H]glucosamine for glycoproteins. Synthetic incorporation of the radiolabel should not result in artifacts during subsequent electrophoresis, although it is theoretically possible that ionizing radiation could damage the proteins.

Proteins can also be radiolabeled post-synthetically, prior to electrophoresis, using a variety of approaches such as iodination with ^{125}I or reductive methylation with [^{3}H]sodium borohydride, reviewed in reference [22]. The disadvantage of this approach is that post-synthetic radiolabeling procedures can readily result in protein charge modifications, leading to artifacts during electrophoresis in the absence of SDS, IEF or 2-D PAGE.

9.7.2 Gel drying

For techniques in which the gel is to be placed in direct contact with X-ray film (i.e. autoradiography and fluorography) it is essential that the gel is first dried. Thin and ultra-thin gels cast on glass or plastic supports can, after equilibration in 3% (w/v) glycerol, be dried simply in air or in an oven at 40–50°C.

Unsupported gels cannot be dried in this way as they will shrivel. The simplest method of drying is to equilibrate the gel in 3% (w/v) glycerol to prevent cracking and then place it between two cellophane sheets supported in a plastic frame. Such frames are commercially available. Drying takes place overnight at room temperature or about 1 h using hot air. Such gels are usually dried down onto a filter paper support, which attaches to the gel during drying and preserves its original dimensions.

The most popular method for drying gels is by heating under vacuum, and suitable apparatus is available from several manufacturers (*Figure 9.5*). Prior to drying, the gel should be soaked for several hours in 3% (w/v) glycerol to prevent cracking and in the presence of 30% (v/v) methanol to minimize swelling. Gradient gels present a particular risk for cracking and swelling: such gels can be satisfactorily dried after soaking in a solution containing 2% (v/v) dimethylsulfoxide (DMSO) and 1% (w/v) glycerol. The gel, supported on a sheet of prewetted 3MM filter paper, is then placed on the drier support beneath which is the heating block. The gel is covered with porous cellophane or Saran Wrap (cling film), then a porous plastic sheet, and finally the silicone rubber cover sheet of the apparatus which forms a vacuum seal. The device is then connected to a vacuum pump, which should be fitted with a cold trap. Gels are usually dried at 80°C, but some equipment allows lower temperatures to be used (e.g. 60°C) which can be helpful in minimizing cracking of a high %T gel.

FIGURE 9.5: *Commercial apparatus for drying polyacrylamide gels. Photograph courtesy of Bio-Rad Laboratories Ltd.*

9.7.3 Autoradiography

Radiolabeled proteins separated by electrophoresis can be detected by direct autoradiography in which the dried gels are placed in contact with an X-ray film and exposed for the appropriate time. Gels can be subjected to autoradiography without drying, but then the film must be protected from the adverse effects of moisture by wrapping the gel in cling film. This method is useful if the gel is to be manipulated after autoradiography (e.g. if radioactive proteins are to be localized prior to elution from the gel) but resolution is generally superior if gels are dried prior to autoradiography (see Section 9.7.2). Direct autoradiography of isotopes such as ^{14}C, ^{35}S, ^{32}P, and ^{125}I is relatively efficient, but ^{3}H is detected very inefficiently as its low energy β-particles do not penetrate the gel matrix and are, therefore, severely quenched. Quenching is also a problem if gels are stained, particularly using silver-based methods, prior to autoradiography.

9.7.4 Fluorography

A more sensitive method of scintillation autoradiography, called fluorography, was developed for the detection of weak β-emitters such as ^{3}H, ^{14}C, and ^{35}S in polyacrylamide gels [23]. In this procedure, the fixed wet gel is impregnated with a scintillant so that low-energy β-particles, which are unable to penetrate through the gel, can excite the fluor molecules to emit photons which can form a photographic image. Most fluors emit blue light so that a blue-sensitive X-ray film should be used.

In the original procedure, 2,5-diphenyloxazole (PPO) was employed as the fluor. The usual solvent for PPO is DMSO, but it is probably safer and less expensive to use acetic acid. Water-soluble fluors, such as sodium salicylate, can be used as inexpensive substitutes for PPO although they do appear to result in increased autoradiographic spreading leading to a loss of resolution [24]. Fluorographic reagents are now also available from commercial suppliers and these are more convenient and safer to use than PPO, but may have a lower sensitivity.

Pre-exposure of the X-ray film to a brief flash of light (approximately 1 msec) greatly increases (two- to threefold) the sensitivity of fluorography [25]. This process also facilitates quantitative analysis since it corrects the non-linear relationship between radioactivity and absorbance of the film image. The use of low temperatures for the fluorographic exposure also results in increased detection sensitivity. Exposure at $-70°C$ results in a 12-fold increase in sensitivity for ^3H and a nine-fold increase for ^{14}C and ^{35}S. The main disadvantage of fluorographic procedures is that they result in more diffuse protein zones compared with direct autoradiography, resulting in effective loss of resolution [25].

9.7.5 Intensification screens

In autoradiography of high-energy β-emitters such as ^{32}P and γ-emitters such as ^{125}I, much of the emission passes directly through the film. It is, therefore, more efficient to place the film between the dried gel and an image intensification screen (e.g. calcium tungstate). The strong emissions which pass through the film can then excite the fluor or phosphor in the screen, thereby creating a secondary fluorographic image superimposed on the primary autoradiographic image. Two screens, arranged in the order gel, screen 1, film, screen 2, can further increase sensitivity of detection of ^{32}P (but not ^{125}I) and this method is often used for visualization of phosphoproteins and nucleic acids. This extra sensitivity is gained at the expense of decreased resolution due to spreading of the zones.

9.7.6 Dual isotope methods

In the approach developed by McConkey [26], two samples to be compared are radiolabeled with different analogs of the same amino acid, for example, ^3H- and ^{14}C- or ^{35}S-labeled methionine. After radiolabeling, the samples are mixed and subjected to co-electrophoresis on the same 1- or 2-D gel. The resultant gel is processed for fluorography which will detect proteins labeled with either isotope (i.e. ^3H or $^{14}C/^{35}S$) and produce a composite image of both samples. Subsequently, the gel is subjected to direct autoradiography to detect only the stronger (i.e. ^{14}C or ^{35}S) emissions. The patterns derived from the proteins of each sample are then interpreted subtractively. The success of this method depends on using the appropriate input ratio of ^3H to ^{35}S (or ^{14}C) radioactivity in the sample mixture applied to the gel. This ratio is dependent on the isotopes used, gel thickness, the fluor used for fluorography, and the X-ray films used for each type of exposure [27].

An alternative method has been described [28] using a combination of [^{35}S]methionine and [^{75}Se]selenomethionine, the latter being a γ-emitter. The gel is again subjected to a fluorographic exposure to detect both isotopes, while a second autoradiographic exposure with an exposed X-ray film interposed between the gel and the film is used to detect only the γ-emitter. However, this method results in some loss of resolution due to the increased zone spreading commonly encountered with γ-emitting isotopes.

The main disadvantage of the dual labeling techniques described above is that two separate films have to be exposed, as black and white emulsions cannot discriminate between decays of different energies. To overcome this problem, color negative film which contains three photographic emulsion layers can be used [29]. Used in autoradiography, the first layer produces a yellow image, the second layer a magenta image, and the third layer a cyan image. A weak β-emitter such as ^{3}H will expose only the upper emulsion producing a yellow image. Stronger β-emitting isotopes (^{14}C and ^{35}S) penetrate to the second layer producing a red image (yellow + magenta), while very energetic β (^{32}P) or γ (^{75}Se, ^{125}I) emitters will expose all three layers, giving a neutral density. This method works best with films designed for tungsten illumination and is also compatible with fluorographic procedures.

9.7.7 Electronic detection methods

Despite their inherent simplicity, techniques based on autoradiography suffer from the disadvantage that long exposure times are often necessary. In addition, problems of non-linearity of film response, limited dynamic range, autoradiographic spreading, and fogging complicate quantitative analysis. To overcome these problems, several electronic methods of detecting radioactive proteins directly in gels have been developed and are reviewed in reference [30]. This topic is discussed in more detail in Chapter 10.

9.7.8 Gel fractionation and counting

Radioactive proteins, located on gels by staining or autoradiography, can be cut out and quantified using either a scintillation counter (β-emitting isotopes) or a gamma counter (γ-emitting isotopes). This method has the added advantage that emissions of different energies can be differentiated so that the results of dual-labeling experiments can be directly quantified. The method is, however, very laborious as the gel must be cut into many small segments if the resolution of the original separation is to be preserved. In addition, for scintillation counting it is necessary to elute the proteins from the gel slices in order to obtain optimal counting efficiency. It is possible to leach the proteins out of the slices using special reagents compatible with the scintillation cocktail. Alternatively, the gel slices can be solubilized, either by heating at 50°C in 30% (w/v) hydrogen peroxide for gels cross-linked with Bis or by using alternative cross-linkers such as DHEBA and DATD (see Section 3.6) which are more easily hydrolyzed.

9.8 Detection of glycoproteins

Glycoproteins will, of course, be detected in gels using the general protein staining methods (see Sections 9.4–9.6), but in addition there are specific detection methods that allow them to be distinguished from non-glycosylated proteins. Until recently, the most widely used method was the periodic acid–Schiff stain using fuchsin [31]. More sensitive variations on this technique have been described using Alcian blue, dansyl hydrazine, thymol-sulfuric acid, or in combination with a modified silver staining technique (see *Table 9.4*); for an extensive bibliography of these techniques see reference [32].

Radiolabeled, fluorescent, or enzyme-conjugated lectins are currently the most versatile reagents for the characterization of glycoproteins separated by electrophoresis. These reagents can be applied directly to the gel matrix, but are better used in conjunction with the technique of Western blotting (see Chapter 11). Methods are also available for radiolabeling the carbohydrate portions of glycoproteins either synthetically or post-synthetically prior to electrophoresis, followed by the detection of the radiolabeled glycoproteins after electrophoretic separation by autoradiography or fluorography.

TABLE 9.4: *Modified silver staining procedure for the detection of glycoproteins and polysaccharides [35]*

Step	Procedure	Conditions
1	Soaking	Overnight in 25% (v/v) isopropanol containing 10% (v/v) acetic acid
2	Incubation	30 min in 7.5% (v/v) acetic acid
3	Incubation	60 min at 4°C in 0.2% (w/v) aqueous periodic acid
4	Rinsing	3 × 10 min in 1 l distilled water
5	Soaking	> 2 h (or overnight) in water
6	Silver staining	1–15 min in silver diamine solution (prepared by adding 1.4 ml concentrated ammonium hydroxide to 21 ml of 0.36% (w/v) sodium hydroxide, stirred vigorously and 4 ml of 19.4% (w/v) silver nitrate slowly added and when the transient precipitate has cleared, made up to 100 ml with water)
7	Rinsing	2 min in water
8	Development	5–30 min in a solution containing 0.05% (w/v) citric acid, 0.019% (w/v) formaldehyde.
9	Rinsing	Thoroughly in water

This method is claimed to be 64 times more sensitive than the periodic acid–Schiff stain and to be able to detect 0.4 ng bound carbohydrate. In this procedure, the treatment with periodic acid is followed by the silver staining method of Oakley *et al.* [8], but other methods may also be applicable (see Section 9.6).

9.9 Detection of phosphoproteins

The most commonly used approach to the analysis of phosphoproteins is to label them *in vivo* with ^{32}P and then to detect them after electrophoresis by autoradiography. Methods are also available for the specific staining of phosphoproteins in gels [32], for example, the reagent Stains-All™ stains phosphoproteins blue while the majority of other proteins are stained red.

9.10 Detection of lipoproteins

Lipoproteins can be stained with Sudan black B or Lipid crimson prior to electrophoresis. Sudan black B can also be used as a post-electrophoretic stain for lipoproteins.

9.11 Detection of enzymes

In order to detect specific enzyme activities in gels it is essential that precautions are taken to minimize loss of enzyme activity during electrophoresis. Denaturing conditions (e.g. urea and SDS) must be avoided, although there are reports of successful renaturation of enzymes after 2-D PAGE: the buffer system and pH for the separation must be chosen with care. Pre-electrophoresis of the gels is also recommended to remove unreacted monomers and catalysts. Gels cannot be fixed prior to visualization of enzyme activity, so that resolution can be degraded if the separated proteins diffuse appreciably during the time taken to carry out the reaction.

Enzymes can be assayed following gel slicing and elution, but this method is very laborious, so specific enzyme staining methods on the intact, unfixed gel are usually employed. Enzyme staining can be accomplished by incubating the gel in a solution of the appropriate reagents using either fluorogenic or chromogenic substrates. This method is quite satisfactory if the final reaction product is insoluble, but a soluble reaction product will rapidly diffuse away from the site of enzyme activity. Better results are generally obtained by partially immobilizing the reagents using a 'contact print' or gel overlay technique. In this approach, the appropriate substrates and other reagents are either impregnated into a filter support (e.g. paper or cellulose acetate) or included in a thin layer of agarose or polyacrylamide gel cast on a glass or plastic support. The overlay is then placed in direct contact with the surface of the separation gel and incubated for an appropriate time, after which the enzymic activity of the separated components is visualized on the overlay. This type of 'print' method can work quite satisfactorily even for soluble reaction products, as their rate of diffusion is significantly reduced.

The majority of enzyme staining methods currently used are based on electron transfer dyes such as methyl thiazolyl tetrazolium, which is reduced by electron donors to form a dark blue insoluble product, formazan. This reaction is catalyzed by phenazine methosulfate and can be used to detect enzymes which lead to the production of NADH or NADPH. This approach can be extended to other enzymes if they can be coupled by intermediate reactions to the reduction of NAD^+ or $NADP^+$. Methods have been developed for the detection of a very large number of enzyme activities, following separation of proteins using a variety of gel systems [32, 33].

References

1. Merril, C.R. (1987) in *Advances in Electrophoresis*, Vol. 1 (A. Chrambach, M.J. Dunn, and B.J. Radola, eds). VCH, Weinheim. 111–139.
2. Neuhoff, V., Stamm, R. and Eibl, H. (1985) *Electrophoresis*, **6**, 427–448.
3. Neuhoff, V., Arold, N., Taube, D. and Ehrhardt, W. (1988) *Electrophoresis*, **9**, 255–262.
4. Barger, B.O., White, F.C., Pace, J.L., Kemper, D.L. and Ragland, W.L. (1976) *Anal. Biochem.*, **70**, 327–335.
5. Jackson, P., Urwin, V.E. and Mackay, C.D. (1988) *Electrophoresis*, **9**, 330–339.
6. Switzer, R.C., Merril, C.R. and Shifrin, S. (1979) *Anal. Biochem.*, **98**, 231–237.
7. Rabilloud, T. (1990) *Electrophoresis*, **11**, 785–794.
8. Oakley, B.R., Kirsch, D.R. and Morris, N.R. (1980) *Anal. Biochem.*, **105**, 361–363.
9. Wray, W., Boulikas, T., Wray, V.P. and Hancock, R. (1981) *Anal. Biochem.*, **118**, 197–203.
10. Hochstrasser, D.F. and Merril, C.R. (1989) *Appl. Theoret. Electrophoresis*, **1**, 35–40.
11. Blum, H., Beier, H. and Gross, H.J. (1987) *Electrophoresis*, **8**, 93–99.
12. Morrissey, J.M. (1981) *Anal. Biochem.*, **117**, 307–310.
13. Rabilloud, T., Carpentier, G. and Tarroux, P. (1988) *Electrophoresis*, **9**, 288–291.
14. Heukeshoven, J. and Dernick, R. (1986) in *Elektrophorese Forum '86* (B.J. Radola, ed.). Technische Universität, München, p. 22–27.
15. Rabilloud, T. (1992) *Electrophoresis*, **13**, 429–439.
16. Sammons, D.W., Adams, L.D. and Nishizawa, E.E. (1982) *Electrophoresis*, **2**, 135–141.
17. Merril, C.R., Harasewych, M.G. and Harrington, M.G. (1986) in *Gel Electrophoresis of Proteins* (M.J. Dunn, ed.). Wright, Bristol, p. 323–362.
18. Dion, A.S. and Pomenti, A.A. (1983) *Anal. Biochem.*, **129**, 490–496.
19. Heukeshoven, J. and Dernick, R. (1985) *Electrophoresis*, **6**, 103–112.
20. Gersten, D.M., Rodriguez, L.V., George, D.G., Johnston, D.A. and Zapolski, E.J. (1991) *Electrophoresis*, **12**, 409–414.
21. Billington, D., Jayson, G.G. and Maltby, P.J. (1992) *Radioisotopes* (J.M. Graham, and D. Billington, eds). BIOS Scientific Publishers, Oxford.

22. Hames, B.D. (1990) in *Gel Electrophoresis of Proteins: a Practical Approach* (B.D. Hames, and D. Rickwood, eds). IRL Press, Oxford, p. 1–147.
23. Bonner, W.M. and Laskey, R.A. (1974) *Eur. J. Biochem.*, **46**, 83–88.
24. Chamberlain, J.P. (1979) *Anal. Biochem.*, **98**, 132–135.
25. Laskey, R.A. (1980) *Meth. Enzymol.*, **65**, 363–371.
26. McConkey, E.H. (1979) *Anal. Biochem.*, **96**, 39–44.
27. McConkey, E.H. and Anderson, C. (1984) *Electrophoresis*, **5**, 230–232.
28. Lecocq, R., Hepburn, A. and Lamy, F. (1982) *Anal. Biochem.*, **127**, 293–299.
29. Kronenberg, L.M. (1979) *Anal. Biochem.*, **93**, 189–195.
30. Sutherland, J.C. (1993) in *Advances in Electrophoresis*, Vol. 6 (A. Chrambach, M.J. Dunn, and B.J. Radola, eds). VCH, Weinheim, in press.
31. Zaccharias, R.J., Zell, T.E., Morrison, J.H. and Woodlock, J.J. (1969) *Anal. Biochem.*, **31**, 148–152.
32. Hames, B.D. (1990) in *Gel Electrophoresis of Proteins: a Practical Approach* (B.D. Hames, and D. Rickwood, eds). IRL Press, Oxford, p. 324–345.
33. Harris, H. and Hopkins, D.A. (1976) *Handbook of Enzyme Electrophoresis in Human Genetics*. North Holland Press, Amsterdam.
34. Hochstrasser, D.F., Patchornik, A. and Merril, C.R. (1989) *Appl. Theoret. Electrophoresis*, **1**, 35–40.
35. Dubray, G. and Bezard, G. (1982) *Anal. Biochem.*, **119**, 325–329.

10 Quantitative Analysis

10.1 Background

In many applications of electrophoretic techniques it is desirable to determine the relative quantities of the separated protein components. Quantitation is not possible by simple visual analysis of the stained separation profiles, so a system is required which can image the electrophoretic profile and subsequently detect and quantitate the individual separated protein zones. One caveat that should be made concerning the quantitation of electrophoretic patterns is that, at best, semi-quantitative data are obtained. Individual proteins can respond quite differently to different staining methods; indeed color development in a staining procedure can vary dramatically with the amino acid composition of the particular protein and can be significantly influenced by the presence of non-protein groups such as carbohydrate. Moreover, there is only a limited range of protein concentration over which there is a linear relationship between color development and concentration. It is a mistake, therefore, to believe that all proteins stain equally with any given detection method and that absolute quantitative data can be obtained.

10.2 Gel imaging

10.2.1 Densitometers

Densitometry is the most widely used method for the imaging of stained gel profiles or autoradiographic images of radiolabeled protein patterns. Several instruments specifically designed for gel scanning are available and attachments are available for fitting to some spectrophotometers. In these systems, the gel is moved mechanically at a constant speed perpendicular to a narrow, fixed, parallel light beam and the amount of light transmitted through the gel is detected by a photomultiplier. Appropriate filters should be used for the optimal detection of the particular staining method (e.g. the peak absorption of

CBB R-250 is in the region 560–575 nm). The resolving capacity of these scanners is variable and usually related to the cost of the instrument. Moreover, the rate of data acquisition by these devices is low, so that they are not generally suitable for application to 2-D gel patterns. Rotating drum densitometers (e.g. Optronics; Joyce-Loebl) capable of scanning film images at high resolution (down to 12.5 µm) and at high speed have been used for the analysis of autoradiographs of 2-D gels [1], but these are very expensive and have largely been replaced by the other imaging devices discussed in the following sections.

10.2.2 Laser densitometers

Several flat-bed scanning densitometers are now available based on the use of a HeNe (633 nm) laser light source. These are high resolution devices (typically down to 50 µm) capable of scanning gels at high speed and with a high dynamic range, so making them applicable to the analysis of 2-D gel profiles. A good example of such a device is the Molecular Dynamics 300 densitometer (*Figure 10.1*) which can scan an image of up to 36 × 42 cm in less than 3 min at a spatial resolution of 88 µm and with a dynamic range of 0.01–4.0 absorbance units. A computer screen image of a 2-D gel digitized with this type of scanner is shown in *Figure 10.2*. One significant disadvantage of these devices is that they are insensitive to objects whose color approaches that of the laser light source. This is a particular problem when using HeNe lasers (red) for silver-stained gels where protein zones are sometimes stained orange to red rather than the optimal dark brown color. It is possible to overcome this problem by copying the silver-stained gel using X-ray duplicating film to produce a film image of black spots on a transparent background.

10.2.3 TV cameras

Both 1-D and 2-D gel patterns can be imaged rapidly with cameras using vidicon tubes. These devices are relatively inexpensive and are

FIGURE 10.1: The gel analysis system in the author's laboratory, comprising a Molecular Dynamics 300A laser densitometer (A) connected by Ethernet to a Sun SparcStation 1 (B) running software for the analysis of 1-D and 2-D separations.

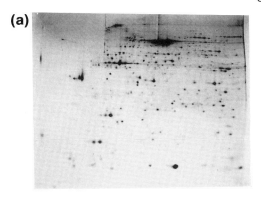
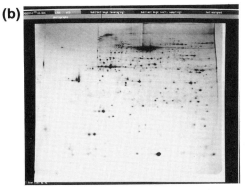
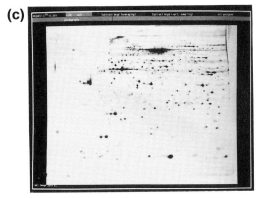

FIGURE 10.2: Initial stages in computer analysis of 2-D gels using the PDQUEST system. (a) 2-D PAGE separation of human heart proteins visualized by silver staining; (b) screen display of the raw digitized image of the gel in (a) acquired with a Molecular Dynamics 300A laser densitometer; (c) image of the gel after filtering and spot subtraction.

compatible with both stained gels and autoradiographic images using either trans- or epi-illumination. However, the dynamic range of these devices (typically 0–1.4 absorbance units) is limited. Spatial resolution is also poor (usually 512 × 512 pixels), although more expensive systems with improved performance are available (1024 × 1024 pixels).

10.2.4 CCD array scanners

These devices are essentially cameras fitted with one- or two-dimensional arrays of photodiodes mounted in the focal plane behind a lens. They allow rapid image acquisition at high dynamic range. Their disadvantages include problems in maintaining normalization of the photodiodes to each other over extended periods of time, lens flare, and correcting for non-uniformity in the background. Cooled CCD cameras, with improved photometric precision and spatial resolution, have also been used [2]. The extreme sensitivity of these devices to low light levels and their ability to build up an image over many minutes makes them applicable to the imaging of fluorescently labeled proteins in gels [2].

10.2.5 Radioisotope imagers

Gel profiles of radiolabeled proteins can be conveniently detected by autoradiography or fluorography using X-ray film (see Sections 9.7.3 and 9.7.4). The film images can then be analyzed quantitatively using one of the methods described above. However, this approach suffers from several disadvantages which complicate quantitative analysis, including: (i) lack of sensitivity, (ii) limited dynamic range, (iii) non-linearity of film response, (iv) autoradiographic spreading, and (v) fogging. To overcome these problems, several electronic methods of detecting radioactive proteins directly in gels have been developed (reviewed in reference [3]). One approach is the use of gas counters based on the multiwire proportional counter and multistep avalanche technologies (*Figure 10.3*) [4–5]. This technique provides an increase in

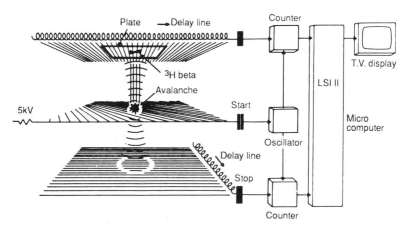

FIGURE 10.3: A schematic representation of the operation of a multiwire proportional counter autoradiographic imaging system which uses the delay line approach to electronic readout of the counter. The electrophoretic plate is placed adjacent to one of the cathode wire planes so that the β-particles can travel into the gas space and trigger an avalanche. There may or may not be a thin window to isolate the sample from the counter gas. Reproduced from reference [4] with permission from VCH.

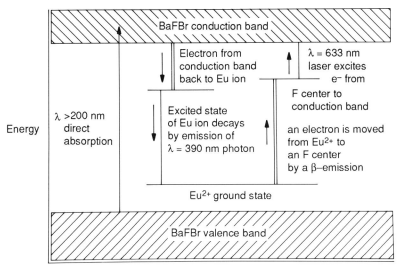

FIGURE 10.4: *Energy level diagram of the states involved in phosphor-imaging using BaFBr:Eu^{2+}.*

sensitivity of 10–20 times compared with autoradiography, coupled with a high dynamic range, but the spatial resolution (typically around 0.5 mm) is poor. Other approaches have been described using a multichannel plate electron multiplier in conjunction with a resistive anode position detector [6], and a low-flux photon image intensifier [7]. However, none of these methods has gained general popularity and they are very expensive.

10.2.6 Photostimulable phosphor-imaging systems

An exciting development based on the use of photostimulable storage phosphor-imaging plates has recently been described [8] and should now be considered the method of choice for quantitative imaging of β-emitting radioisotopes [3]. These imaging plates, commercially available from Fuji or Kodak, are composed of a thin (approximately 500 μm) layer of very small crystals of BaFBr:Eu^{2+} in a plastic binder. Like other insulators, BaFBr crystals have a valence 'band' of closely spaced energy states, all of which are occupied by electrons that move freely through the crystal, and a conduction band of unoccupied energy states. Passage of a β-particle through a crystal results in the transfer of an electron from the Eu^{2+} ion to a localized energy state at a site in the crystal where a Br$^-$ ion is missing, that is, a color or 'F' center (*Figure 10.4*). This process converts the Eu^{2+} to Eu^{3+}. Electrons are trapped in F centers which are metastable. Photons (i.e. light) of wavelengths less than 800 nm have sufficient energy to transfer an electron from the F center into the conduction band, from which it can

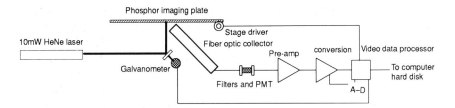

FIGURE 10.5*: Diagrammatic representation of the Molecular Dynamics 400A Phosphorimager. Reproduced from reference [8] with permission from VCH.*

recombine with a Eu^{3+} ion, which is formed in an excited state. The excited state of the Eu^{3+} returns to its ground state by emitting a photon of a wavelength of 390 nm.

In practice, a phosphor-imaging plate is exposed to the dried 1-D or 2-D gel containing separated radiolabeled protein zones. After the plate is exposed, it is transferred to a scanner where light from a HeNe laser (633 nm) is absorbed by the F centers, resulting in the emission of a blue (390 nm) luminescence proportional to the original amount of radiation incident on the plate (*Figure 10.5*). Several commercial instruments exploiting this technology (e.g. Molecular Dynamics 400A PhosphorImager, Fuji BAS 1000, Bio-Rad Molecular Imaging System) are available (*Figure 10.6*). These systems require relatively short exposure times, as a phosphor plate can capture the image of a highly labeled gel in about 10% of the time required for conventional autoradiography, and have a very high dynamic range. However, the commercial instruments are very expensive.

FIGURE 10.6*: Bio-Rad Model GS-250 Molecular Imager system using storage phosphor plate technology for rapid direct imaging and quantitation of radioactive and chemiluminescent samples. Photograph courtesy of Bio-Rad Laboratories.*

Photostimulable phosphor-imaging systems can also be used in dual labeling experiments involving two radioisotopes with energies that differ substantially, for example ^{35}S and ^{32}P [9]. Two images are recorded, one with the imaging plate in direct contact with the gel and the other with a thin metal foil interposed between them. A 36 μm copper foil attenuates β-emissions from ^{35}S 750-fold, while reducing β-emissions from ^{32}P by only 30% [9]. Detection efficiencies are determined in control experiments and used to discriminate the two experimental images in order to determine the distribution of each radioisotope in the gel. This method has been successfully applied to the simultaneous analysis of phosphoproteins and total cellular proteins of PC12 cells [10].

10.3 Analysis of one-dimensional gels

For the quantitative analysis of one-dimensional gel patterns, the output from the image acquisition device is usually fed to a chart recorder (*Figure 10.7*) and the individual protein components are quantified by measurement of the areas under the individual peaks. This is most conveniently achieved using an automated integrator, but if such a facility is not available it is possible to cut out the appropriate areas and weigh them. Some investigators prefer to measure peak

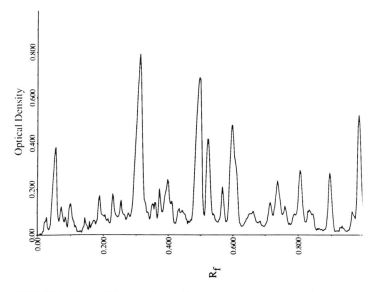

FIGURE 10.7: *Printout from the computer screen of a densitometric trace, obtained with the Molecular Dynamics 300A laser densitometer, of a 10%T SDS–PAGE separation of human heart proteins shown in* Figure 6.1 *(lane c).*

height, but this method is less accurate as peak height depends not only on the amount of material present, but also on the distance travelled through the gel. Increasingly, commercial densitometers are being fitted with microcomputer work-stations running software that allows automated logging and quantitative analysis of the data.

10.4 Analysis of two-dimensional gels

High-resolution 2-D PAGE (see Chapter 8) results in highly complex two-dimensional protein maps from which only a limited amount of information can be extracted by simple visual inspection. The analysis and interpretation of this mass of information requires automated computer analysis systems which can: (i) extract both qualitative and quantitative information from individual gels, (ii) provide pattern matching between gels, and (iii) allow the construction of databases for different types of sample. Only a flavor of this specialized topic can be provided here and the reader requiring further information on this area is referred to some recent review articles [1,11,12].

If proper quantitative analysis of 2-D gels is to be performed, it is essential that the image acquisition device is accurately calibrated. For stained gels this is usually achieved by digitization of optical density standards; for radioactive samples by using calibration strips of acrylamide containing known amounts of radioactivity. These standards are then used to construct a calibration curve which converts the density value into an amount of protein.

The first step in the analysis of 2-D gels is usually a filtering procedure to reduce image noise. A process of background subtraction then follows, which is not a trivial task, as the background values across a 2-D gel can vary considerably. This precludes the use of a simple threshold procedure so that more sophisticated algorithms must be employed. In many of the software packages, streaks are also removed from the 2-D images on the basis that, although they represent sample protein, they cannot be attributed to a particular spot and are thus not amenable to analysis. The effect of these 'clean-up' procedures on a typical 2-D gel image is shown in *Figure 10.2*.

In the next step of the analysis, the individual spots making up the 2-D gel pattern must be detected and resolved. This task is complicated, as not all spots are resolved as discrete entities, many being contiguous or overlapping. Different methods have been devised to solve this problem, including thresholding, neighborhood transformation, contrast enhancement, and 2-D Gaussian fitting [1,11]. Once the individual protein spots have been detected, it is then a relatively simple task to determine the total density of each spot and, using the calibration information, convert this to protein amount.

At this stage, quantitative analysis of the individual 2-D gel patterns is complete. However, most investigators require comparative analysis of a series of 2-D gels. This requires software for the matching of 2-D gel patterns. This is not a trivial process, due to considerable spatial variability which can occur between gels, so that sophisticated algorithms (e.g. nearest-neighbor analysis, rubber sheet transformation, least-square fitting) are used to bring the global images of individual 2-D gels into coincidence [1,11].

The final level of complexity is provided by software that is able to cross-match multiple sets of 2-D gels [11, 12] to provide large-scale databases of 2-D gel patterns [12] which can be readily interfaced to other databases such as those containing protein and DNA sequences.

Several laboratories and commercial companies have developed computer systems for automated quantitative analysis of 2-D gels. Until recently, these systems required the use of large and expensive mainframe or mini-computers, specialized ancillary hardware such as array processors, and even custom-built hardware [1]. Fortunately, recent developments resulting in powerful, yet affordable, microcomputer work-stations (e.g. Sun, VAX) have produced systems which are more readily available to the scientific community (*Figure 10.1*). Some of the commercially available systems are listed in *Table 10.1*. The larger systems (e.g. PDQUEST, Kepler, BioImage) provide a full 2-D gel analysis capability including scanning, spot extraction, quantitation, matching, database construction, and analysis tools. Systems

TABLE 10.1: *Commercial computer software packages for analysis of 2-D gels*

System name	Derived from	Computer system	OS (operating system)	Supplier
Visage	–	SUN	UNIX	Millipore
PDQUEST	QUEST	SUN	UNIX	Protein and DNA Imageware Systems
Kepler	TYCHO	VAX	VMS	Large Scale Biology Corporation
Biolog	HERMeS	SUN	UNIX	Biolog
Gemini	–	Custom	Unknown	Applied Imaging
IB-1000	–	IBM-AT	MS-DOS	Indiana Biotech
Microscan 1000	–	IBM-AT	MS-DOS	Technology Resources Inc
Phoretix-1	–	IBM-AT	MS-DOS	Biometra Ltd
QGEL/QBASE	–	IBM-AT	MS-DOS	Quanti-Gel Corporation

For supplier's addresses, see Appendix B.
Reproduced from reference [11] with permission from Oxford University Press.

based on desktop personal microcomputers (e.g. IBM AT) are at present limited in their capabilities, especially in terms of their ability to match large sets of 2-D gel patterns. However, a cut-down version of PDQUEST, known as ImageMaster (Pharmacia–LKB), has recently been implemented on PC workstations. This system is able to simultaneously identify and match up to six 2-D gels, each with up to 10 000 protein spots.

References

1. Miller, M.J. (1989) in *Advances in Electrophoresis*, Vol. 3 (A. Chrambach, M.J. Dunn and B.J. Radola, eds). VCH, Weinheim, p. 181–217.
2. Jackson, P., Urwin, V.E. and Mackay, C.D. (1988) *Electrophoresis*, **9**, 330–339.
3. Sutherland, J.C. (1993) in *Advances in Electrophoresis* (A. Chrambach, M.J. Dunn and B.J. Radola, eds). VCH, Weinheim, in press.
4. Bateman, J.E. (1990) *Electrophoresis*, **11**, 367–375.
5. Charpak, G., Dominik, W. and Zaganidis, N. (1989) *Proc. Natl Acad. Sci. USA*, **86**, 1741–1745.
6. Burbeck, S. (1983) *Electrophoresis*, **4**, 127–133.
7. Davidson, J.B. (1984) in *Electrophoresis '84* (V. Neuhoff ed.). Verlag Chemie, Weinheim, p. 235–251.
8. Johnston, R.F., Pickett, S.C. and Barker, D.L. (1990) *Electrophoresis*, **11**, 355–360.
9. Johnston, R.F., Pickett, S.C. and Barker, D.L. (1991) *Methods. A Companion to Methods in Enzymology*, Vol. 3 (M.G. Harrington, ed.). Academic Press, San Diego, p. 128–134.
10. Harrington, M.G., Hood, L. and Puckett, C. (1991) *Methods. A Companion to Methods in Enzymology*, Vol. 3 (M.G. Harrington, ed.). Academic Press, San Diego, p. 135–141.
11. Dunn, M.J. (1992) in *Microcomputers in Biochemistry. A Practical Approach* (C.F.A. Bryce, ed.). IRL Press, Oxford, p. 215–242.
12. Celis, J.E., Madsen, P., Gesser, B., *et al.* (1989) in *Advances in Electrophoresis*, Vol. 3 (A. Chrambach, M.J. Dunn, and B.J. Radola, eds). VCH, Weinheim, p. 1–179.

11 Western Blotting

11.1 Background

It should be clear from the preceding chapters that techniques of gel electrophoresis have an almost unrivalled ability to separate the components of complex protein mixtures. Using appropriate methods, sample proteins can be characterized in terms of their mobility, charge, size, and abundance. However, these methods do not provide directly any information on the identity or functional properties of the separated proteins. We have already seen in Chapter 9 how specific staining procedures can be used to identify particular classes of polypeptides and how the enzymic properties of proteins separated by electrophoresis can be examined. In Chapter 12 methods for recovering proteins from gels and approaches to their chemical characterization will be discussed. There are, however, techniques which allow the identity of separated proteins on gels to be probed more directly; these are the subject of this chapter.

The high specificity and affinity of ligands such as antibodies (both polyclonal and monoclonal) and lectins makes these reagents highly sensitive tools for the identification and characterization of proteins separated by gel electrophoresis. Protein samples can be reacted with an appropriate specific antiserum prior to electrophoresis; the resulting immunoprecipitate can then be recovered and the reactive proteins analyzed by electrophoresis. Alternatively the sample, after removal of the immunoprecipitate, can be analyzed to determine which proteins have been depleted or removed. This technique is known as immunodeletion.

It is also possible to use antibodies to precipitate protein components following electrophoretic separation. This technique, known as immunofixation, works well with cellulose actetate membranes and agarose gels and can give satisfactory results with polyacrylamide gels of low %T, as used for IEF. However, the restrictive nature (i.e. the small pore size) of most polyacrylamide gels makes this technique

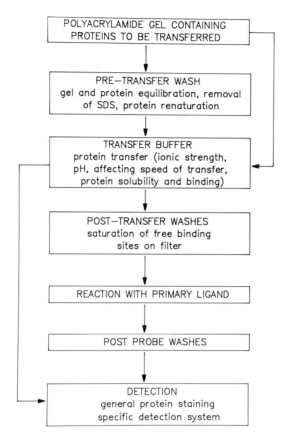

FIGURE 11.1: Schematic outline of the steps involved in Western blotting.

generally unsatisfactory. This is due both to the slow rate of diffusion of antibodies or other large probe molecules into the gel matrix and to the diffusion of the separated protein zones that occurs during probe penetration.

The problems associated with immunofixation were overcome by the development of blotting techniques, in which the separated proteins are transferred from the gel onto the surface of a thin matrix such as nitrocellulose. As a result, the proteins are immobilized on the matrix and are readily accessible to interaction with antibodies and other ligands. This procedure was based on techniques developed by Southern [1] for transfer of DNA (Southern blotting) and subsequently adapted for RNA transfer (Northern blotting). Almost inevitably, when this technique was subsequently adapted to proteins [2] it became popularly known as Western blotting.

The major steps in Western blotting are shown in *Figure 11.1*. After electrophoresis is complete, the gel should be incubated in the appropriate transfer buffer to remove SDS (if SDS–PAGE has been used) and other gel constitutents which can cause problems during transfer. This also minimizes shrinking or swelling of the gel during transfer and allows possible protein renaturation (if this can occur). After transfer by the chosen procedure, the protein pattern can be visualized using a total protein stain. For specific detection, non-specific ligand binding sites on the filter must be blocked before probing with a specific antibody or lectin. Proteins reactive with the probe are then visualized using an appropriate reporter molecule. These steps are discussed in the following sections, but for a more comprehensive overview of Western blotting see reference [3].

11.2 Transfer methods

Protein transfer can be achieved by contact diffusion, capillary or vacuum methods [3]. More rapid and efficient transfer from polyacrylamide gels can generally be achieved by transferring the

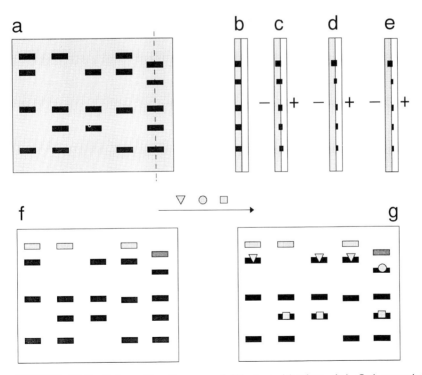

FIGURE 11.2: *Schematic diagram of Western blotting. **(a)** Gel containing separated proteins. **(b–e)** Progressive electrophoretic transfer of protein from the gel onto the surface of the blotting membrane **(f)**. **(g)** Interaction of protein bands with specific ligands.*

separated proteins from the gel onto the surface of a thin matrix by application of an electric field perpendicular to the plane of the gel (*Figure 11.2*). This technique, which was pioneered by Towbin [2], is known as electroblotting.

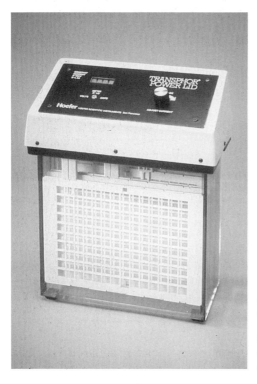

FIGURE 11.3*: Vertical tank apparatus for Western blotting with power supply built into the lid. Photograph courtesy of Hoefer Scientific Instruments.*

A wide variety of electroblotting apparatus has been described and many systems are available commercially; however, there are two basic designs. The first configuration uses vertical buffer tanks (*Figure 11.3*). In this type of blotter the gel and the blotting membrane are held in intimate contact by a sandwich of filter papers and sponge pads held under pressure in a frame fixed between the electrodes (*Figure 11.4*). The design of the electrode assemblies, usually arrays of platinum wires, is important since the rapid, efficient and even transfer of proteins depends on the generation of a homogeneous electrical field over the whole gel area combined with a high current. The electrode assemblies are usually several centimeters apart so that the maximum voltage gradient that can be applied is limited (typically 5 V/cm), even when using an efficient cooling system. Consequently, transfer times of several hours are generally used, but this can be reduced if the proteins to be transferred are of low M_r or if the gels are of a low %T. High current settings of 0.5–1 A are typically used, so that it is essential to have a good power supply and to control Joule heating with a refrigerated recirculator.

WESTERN BLOTTING   143

FIGURE 11.4: Vertical tank apparatus for Western blotting showing plastic frames and sponge pads for the construction of gel-blotting membrane assemblies. Photograph courtesy of Hoefer Scientific Instruments.

The second configuration, known as semi-dry blotting, employs flat-plate electrodes arranged in a horizontal apparatus, providing a homogeneous electrical field, with a short inter-electrode distance and using small amounts of buffer (*Figure 11.5*). The gel and matrix are sandwiched between two stacks of filter papers, wetted with transfer buffer, which are in direct contact with the plate electrodes (*Figure 11.6*). Much higher field strengths at lower current settings can be achieved with this type

FIGURE 11.5: Horizontal semi-dry equipment for Western blotting. The apparatus shown here employs two screen electrodes (one platinum and one stainless steel) and the two versions are for use with mini-gels and standard size gels. Photograph courtesy of Hoefer Scientific Instruments.

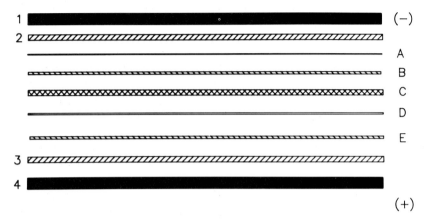

FIGURE 11.6: *Assembly of one gel–membrane assembly (TRANS-UNIT) in a semi-dry blotting apparatus using a discontinuous buffer system [4]. (1,4) Graphite electrodes; (2) six layers of filter paper soaked in 0.04 M 6-amino-n-hexanoic acid, 0.025 M Tris, 20% (v/v) methanol, pH 9.4; (3) six layers of filter paper soaked in 0.3 M Tris, 20% (v/v) methanol, pH 10.4, to neutralize protons produced at the anode during electrophoretic transfer. The TRANS-UNIT: (A) dialysis membrane; (B) three layers of filter paper soaked in 0.04 M 6-amino-n-hexanoic acid, 0.025 M Tris, 20% (w/v) methanol, pH 9.4; (C) polyacrylamide gel; (D) nitrocellulose membrane; (E) three layers of filter paper soaked in 0.025 M Tris, 20% (v/v) methanol, pH 10.4.*

of apparatus. A maximum current of 0.8 mA/cm^2 of gel surface area is usually recommended, so that current settings of around 200 mA are typical. General electrophoresis power supplies are, therefore, suitable and cooling is not usually required. Routinely, blotting is restricted to 2 h or less; again the time depends on the M_r of the proteins and the porosity of the gel. Indeed, prolonged blotting times cannot be used with semi-dry blotters due to problems of evaporation of the limited amount of buffer present.

Several gels can be blotted simultaneously in most tank blotting apparatus. This is also possible in semi-dry designs using stacks of gel–membrane–filter assemblies; it is essential, however, that the individual gel–matrix stacks are separated with dialysis membrane (*Figure 11.6*), to prevent proteins being transferred from one stack to the next [4].

11.3 Blotting matrices

Several types of filters with different properties are available for Western blotting of proteins. Nitrocellulose, the original support described for electroblotting of proteins [2], remains the most popular matrix as it does not require derivatization prior to use, it is compatible with most general protein stains, it is relatively inexpensive and has a high protein binding capacity (249 µg/cm^2 [5]). Standard nitrocellulose has a pore size of 0.45 µm, but membranes of smaller pore sizes are available (0.1 and 0.2 µm) and can give better retention of small proteins ($M_r < 1500$). Nitrocellulose, however, is rather brittle and cracks readily if folded when dry. Several manufacturers now supply nitrocellulose in a backed form which is more robust.

Proteins are not covalently bound to nitrocellulose membranes and are thought to adsorb by a combination of hydrophobic and electrostatic interactions. However, proteins are bound covalently to diazo papers, such as diazobenzyloxymethyl cellulose and diazophenylthioether cellulose, but these papers are seldom used as they must be activated immediately prior to use. Positively charged nylon-based membranes are robust and have a high binding capacity, but have not proved to be popular for Western blotting applications as blocking of unoccupied binding sites is difficult and most general protein stains (i.e. anionic dyes) will react with the matrix. More recently hydrophobic polyvinylidene difluoride (PVDF) membranes have become available [5]. PVDF membranes have high mechanical strength, a protein binding capacity similar to that of nitrocellulose (172 µg/cm^2 [5]) and are compatible with most Western blotting protocols.

Recently considerable interest has developed in the use of Western blotting as a method to isolate proteins for chemical characterization (i.e. amino acid analysis, peptide mapping, N-terminal and internal microsequence analysis). This has required the development of alternative chemically resistant matrices (e.g. glass fiber, PVDF, polypropylene). This topic is discussed in Chapter 12.

11.4 Transfer buffers

The choice of transfer buffer is important as it influences both the efficiency of elution from the polyacrylamide matrix and retention by the blotting matrix. Elution is determined by buffer composition and pH, while protein retention can be influenced by additives such as detergents and methanol. The most commonly used buffer system for electroblotting of proteins from SDS–PAGE gels is that originally described by Towbin *et al.* [2]: 25 mM Tris, 192 mM glycine, pH 8.3. Methanol (10–20%, v/v) is often added to this buffer as it removes SDS from protein–SDS complexes and increases the affinity of binding of proteins to nitrocellulose. However, it should be remembered that methanol acts as a fixative and reduces the efficiency of protein elution, so that extended transfer times (up to 24 h) must be used. This effect is worse for high M_r proteins, so that methanol is best avoided if proteins of $M_r > 100 \times 10^3$ are to be transferred. It was originally claimed that a discontinuous buffer system (*Figure 11.6*) needed to be used in conjunction with semi-dry electroblotting apparatus [4], but it has subsequently been shown that comparable efficiencies are obtained using standard continuous blotting buffer systems [6]. If basic proteins are to be transferred or if acid-urea gels have been used, a transfer solution of 0.7% (v/v) acetic acid is recommended and the proteins are transferred as cations. Several alternative buffer systems, not containing glycine, can be used when the blotted proteins are to be characterized chemically (see Chapter 12).

11.5 General protein staining

A variety of methods have been described for the visualization on the membrane of the total protein pattern following blotting. The sensitivities and applicability of some of these methods are given in *Table 11.1* and some examples of stained blots are shown in *Figure 11.7*. Fast green and Ponceau stains are relatively insensitive, but can be easily removed from proteins after detection to allow for subsequent immunoprobing or chemical characterization. Amido black (0.1% (w/v) in 45% (v/v) methanol, 7% (v/v) acetic acid) is the most commonly used stain for the rapid (staining time, 10 min; destaining time, 5–10 min) and relatively sensitive visualization of proteins on nitrocellulose. CBB R-250 gives a very high background on nitrocellulose, but gives excellent results with PVDF membranes. More sensitive staining can be achieved by staining with India ink (0.1% (v/v) in phosphate-buffered saline (PBS) containing 0.3% (w/v) Tween 20) [7] or colloidal gold particles (Aurodye™) [8], but these procedures are more protracted (several hours to overnight). Nylon membranes are more

TABLE 11.1: *General protein stains for blot transfers arranged in order of increasing sensitivity*

Detection reagent	Approximate sensitivity	Matrix
Fast green FC	–	NC, PVDF
Ponceau S	–	NC, PVDF
Coomassie brilliant blue R-250	1.5 µg	PVDF
Amido black 10B	1.5 µg	NC, PVDF
India ink	100 ng	NC, PVDF
Colloidal iron	30 ng	NC,N, PVDF
In situ biotinylation + HRP-avidin	30 ng	NC,N, PVDF
Colloidal gold	4 ng	NC, PVDF

NC, nitrocellulose; PVDF, polyvinylidene difluoride; N, nylon; HRP, horseradish peroxidase.

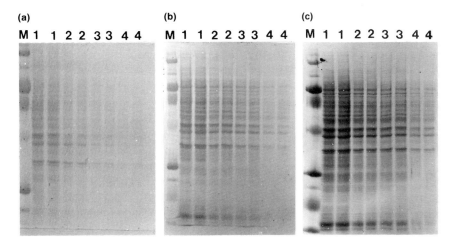

FIGURE 11.7: Western blot transfers of human endothelial cell proteins separated by 10%T SDS–PAGE stained with **(a)** Ponceau (nitrocellulose), **(b)** Amido black (nitrocellulose), and **(c)** Coomassie brilliant blue R-250 (PVDF). Different amounts of protein were applied to each lane: (1) 40 µg, (2) 20 µg, (3) 10 µg, (4) 5 µg, (M) M_r marker proteins.

difficult to stain, but this is possible using a colloidal iron procedure [9] or a method based on *in situ* biotinylation of the blotted proteins followed by visualization with peroxidase-conjugated avidin.

11.6 Protein standards

Protein standard mixtures which have been applied to the gels to provide calibration in terms of M_r or pI will, of course, be visualized along with the sample proteins when general protein stains are used. In addition, colored, pre-stained or biotinylated sets of marker proteins are now available and can be used to provide a visual check of the efficiency of protein transfer. However, it should be realized that derivatization of these proteins can result in changes in molecular mass and/or charge, so that they should be used for gel calibration with caution.

11.7 Blocking

After the proteins have been transferred to the membrane and before probing with a specific ligand, all unoccupied binding sites on the support must be blocked. Bovine serum albumin (3–5% (w/v) in PBS for 1 h) is the most commonly used blocking agent, although other proteins including other animal sera, ovalbumin, hemoglobin, casein, and gelatin have been used. However, many protein preparations used for blocking may be reactive during the subsequent probing, resulting in a high background. A solution of non-fat dried milk (3% (w/v) in PBS) has become popular as a blocking agent and usually results in very low background staining. Another popular blocking agent is the non-ionic detergent, polyethylene sorbitan monolaurate (Tween 20) [10], which should be used at a concentration of 0.1% (w/v) in PBS for 1 h. However, there is a risk of displacing proteins from the blotting matrix if detergents are used for blocking. It should be noted that if nylon membranes are used, these require more vigorous blocking procedures (e.g. 10% (w/v) bovine serum albumin for 12–16 h at 50°C) [3].

11.8 Specific detection

The most important step in the blotting procedure is the reaction with a specific antibody or ligand (e.g. lectin) and visualization of the proteins in the sample with which it reacts. Although it is possible to label the primary antibody (or other ligand) with a suitable reporter group to give direct visualization of reactive proteins on the blot, this approach is not popular as it requires derivatization of the primary antibody which can adversely affect its specificity or affinity. Therefore, it is usual to use an indirect, or sandwich, approach utilizing labeled secondary (or tertiary) antibody reagents.

Following blocking, the membrane is incubated with a solution containing the appropriately diluted specific primary antibody. The membrane is then washed and reacted with a solution containing a secondary antibody specific for the species and immunoglobulin class of the primary antibody. The secondary antibody can be fluorescently labeled (e.g. fluorescein isothiocyanate, FITC), radiolabeled (usually with ^{125}I) or conjugated to an enzyme (e.g. horseradish peroxidase, alkaline phosphatase, β-galactosidase or glucose oxidase). Alternatively, the secondary antibody can be replaced by appropriately labeled staphylococcal protein A or streptococcal protein G. These reagents bind specifically to the F_c region of immunoglobulins (Ig), with protein G reacting with a broader range of species and classes of Ig than does protein A. The sensitivities of some of the commonly used detection systems are shown in *Table 11.2*.

Methods employing fluorescently labeled secondary antibodies require the use of u.v. illumination, while radiolabeled antibody procedures depend on a time-consuming autoradiographic step. Thus, methods based on the use of enzyme-conjugated secondary antibody reagents have become the most popular and kits of these reagents are commercially available. Detection is achieved using appropriate substrates which form insoluble, stable colored reaction products at the sites on the blot where the enzyme-conjugated secondary antibody is bound. The two most popular substrates for use with peroxidase-conjugated antibodies are diaminobenzidine (DAB), which gives brown bands, and 4-chloro-1-naphthol, which produces purple bands. The best substrate system for the visualization of alkaline phosphatase-conjugated secondary antibodies is a mixture of 5-bromo-4-chloro indoxyl phosphate (BCIP) and nitroblue tetrazolium (NBT).

TABLE 11.2: Specific detection methods for blot transfers arranged in order of increasing sensitivity

Method	Approximate sensitivity (ng/mm^2)
Peroxidase–protein A	2.0
Peroxidase–second antibody	1.5
Gold–second antibody	1.5
^{125}I–second antibody	1.0
Peroxidase double sandwich	0.8
Avidin–biotin–peroxidase (ABC) complex	0.5
Gold–second antibody + silver enhancement	0.1
Enhanced chemiluminescence	0.001

Other detection systems have been developed in order to increase the sensitivity of detection of proteins on blots. One approach is the use of colloidal gold-labeled antibodies [11] or protein A. These reagents have the advantage that the stain is visible, due to its reddish-pink color, without further development and the sensitivity of the method can be further increased by silver enhancement. Triple antibody probing methods can also be used to increase the sensitivity of detection. In this case, preformed complexes of enzymes and antibodies are linked by second antibodies to the primary antibody, examples being the peroxidase–antiperoxidase (PAP) and alkaline phosphatase–antialkaline phosphatase (APAP) procedures. Secondary antibodies can also be conjugated with the steroid hapten, digoxigenin, which can then be detected using an enzyme-conjugated anti-digoxigenin antibody [12].

Another popular method of increasing the sensitivity of detection on blots exploits the specificity of the reaction between the vitamin, biotin ($M_r = 224$), and the protein, avidin. Antibodies can be readily conjugated with biotin and the resulting conjugates used as the secondary reagents for probing blot transfers. A third step must then be used for visualization using avidin conjugated with a suitable reporter enzyme (e.g. peroxidase, β-galactosidase, alkaline phosphatase, or glucose oxidase). Even greater sensitivity can be achieved at this stage using preformed complexes of a biotinylated enzyme with avidin, as many enzyme molecules are present in these complexes, giving rise to an enhanced signal. Egg white avidin ($M_r = 68 \times 10^3$) is often used in these procedures, but has the disadvantages that it is highly charged at neutral pH (and so can bind to proteins non-specifically) and is a glycoprotein, which can interact with other biomolecules such as lectins via its carbohydrate moiety. For these reasons it is often preferable to use streptavidin ($M_r = 60 \times 10^3$) isolated from *Streptomyces avidinii* which has a pI close to neutrality and is not glycosylated.

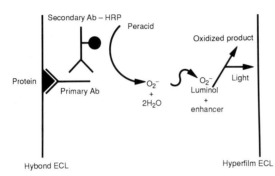

FIGURE 11.8: Diagram showing the mechanism of detection of proteins on Western blots using the ECL system.

The search for increased sensitivity of visualization of immunoblots has resulted in the development of detection systems based on chemiluminescence. The high sensitivity of these methods (they can detect <1 pg of protein) makes it possible to conserve expensive and rare antibodies and other ligands by using them at very high dilutions. Systems are available for use with peroxidase-conjugated and alkaline phosphatase-conjugated secondary antibodies. The procedure for use with peroxidase-conjugated antibodies is known as enhanced chemiluminescence (ECL) and is available from Amersham International [13]. In this system, peroxidase is used to oxidize luminol in the presence of hydrogen peroxide (*Figure 11.8*). Following this reaction luminol is in an excited state and can decay to the ground state via the emission of light. A phenolic enhancer is included in the reaction mixture which is able to enhance the intensity of the light emission by up to 1000 times. The light emission, which peaks after 1–5 min and then decays slowly, is detected by placing the blot in contact with an X-ray film of the appropriate sensitivity. An example of immunodetection using the ECL system is shown in *Figure 11.9*. An additional advantage of this technique is that primary and secondary antibodies can be completely removed ('stripped') after immunodetection and the blot reprobed several times with different primary antibodies.

The system for use with alkaline phosphatase-conjugated antibodies is based on the use of a chemiluminescent dioxetane substrate, disodium 3-(4-methoxyspiro[1,2-dioxetane-3-2'-tricyclo-[3.3.1.1^{3,7}]decan]-4-yl)phenyl phosphate (AMPPD) (*Figure 11.10*) [14]. This reagent is available

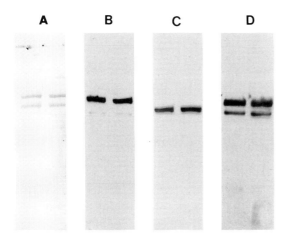

FIGURE 11.9: Result of an immunoblotting experiment to detect the α- and β-subunits of spectrin after SDS–PAGE separation of human erythrocyte membrane proteins. (A) Amido black 10B stained blot; (B–D) visualization using the ECL system. Monoclonal antibodies to: (B) α-spectrin (diluted 1:10,000); (C) β-spectrin (diluted 1:1000); (D) αβ-spectrin (diluted 1:500). The secondary antibody reagent was peroxidase-conjugated rabbit immunoglobulins to mouse immunoglobulins (diluted 1:10 000).

under different trade names from several suppliers (Tropix, Boehringer Mannheim, United States Biochemical, Bio-Rad). The reaction of alkaline phosphatase with AMPPD (*Figure 11.10*) results in dephosphorylation of the molecule and simultaneous destabilization accompanied by the emission of light at 477 nm which can be detected using standard X-ray film or black and white instant photographic film. An

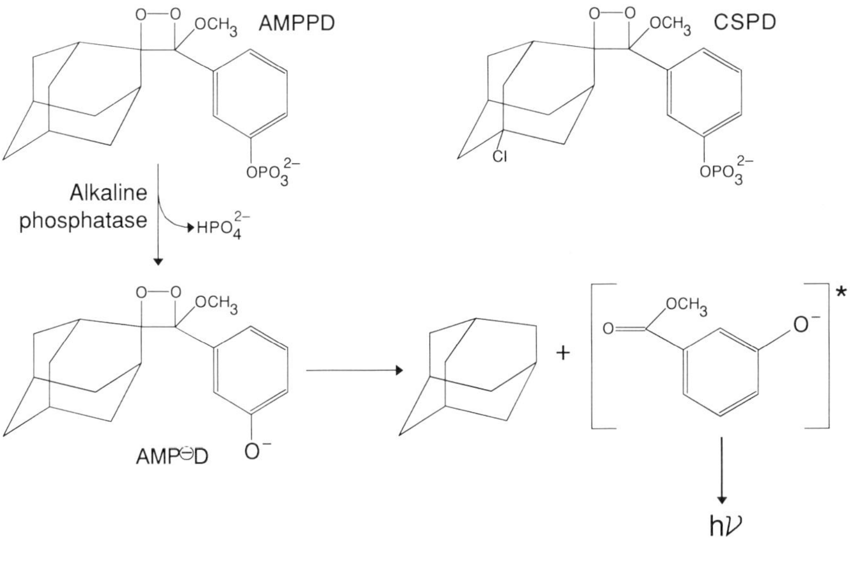

FIGURE 11.10: Structures of AMPPD and CSPD and the mechanism of alkaline-phosphatase-catalyzed chemiluminescence using AMPPD.

enhancer is also available for use with this system and the 5-chloro derivative, CSPD (*Figure 11.10*), produces faster and more sensitive results [14].

11.9 Quantitation

Blotting is generally used as a qualitative technique. If it is to be used as a quantitative procedure, it is essential to determine the transfer efficiencies of the proteins being blotted. This is best accomplished using radiolabeled proteins, so that after transfer, protein bands of: (i) the blot, (ii) the post-blot gel, and (iii) a parallel unblotted gel can be excised and counted or assessed by densitometry of an autoradiograph. In addition, and perhaps more problematically, it must be established that the probe interacts equivalently with all target proteins [3]. The quantitation of reactive proteins can be carried out by autoradiography and densitometry if a radiolabeled probe was used or by reflectance densitometry of colored bands resulting from an enzyme-linked detection system. Transmission densitometry can be used with chemiluminescent detection systems as the results are in the form of photographic films.

References

1. Southern, E.M. (1975) *J. Molec. Biol.*, **98**, 503–517.
2. Towbin, H., Staehelin, T. and Gordon, J. (1979) *Proc. Natl Acad. Sci.*, **76**, 4350–4354.
3. Baldo, B.A. and Tovey, E.R. (1989) *Protein Blotting. Methodology, Research and Diagnostic Applications.* Karger, Basel.
4. Khyse-Anderson, J. (1984) *J. Biochem. Biophys. Methods*, **10**, 203–209.
5. Pluskal, M.G., Przekop, B., Kavonian, M.R., Vecoli, C. and Hicks, D.A. (1986) *BioTechniques*, **4**, 272–283.
6. Bjerrum, O.J. and Schafer-Nielsen, C. (1986) in *Electrophoresis '86* (M.J. Dunn, ed.). VCH, Weinheim, p. 315–327.
7. Hancock, K. and Tsang, V.C.W. (1983) *Anal. Biochem.*, **133**, 157–162.
8. Moermans, M., Daneels, G. and De Mey, J. (1985) *Anal. Biochem.*, **145**, 315–321.
9. Moermans, M., De Raeymaeker, B., Daneels, G. and De May, J. (1986) *Anal. Biochem.*, **153**, 18–22.
10. Batteiger, B., Newhall, W.J. and Jones, R.B. (1982) *J. Immunol. Methods*, **55**, 297–307.
11. Daneels, G., Moermans, M., De Raeymaeker, M. and De Mey, J. (1986) *J. Immunol. Methods*, **89**, 89–91.
12. Kessler, C. (1991) *Molec. Cell. Probes*, **5**, 161–205.
13. Cunningham, M.W., Durrant, I., Fowler, S.J., Guilford, J.A., Moore, M. and Macdonald, R.M. (1992) *International Laboratory*, **22**, 36–40.
14. Bronstein, I., Voyta, J.C., Murphy, O.J., Bresnick, L. and Kricka, L.J. (1992) *BioTechniques*, **12**, 748–753.

12 Chemical Characterization of Proteins Separated by Electrophoresis

12.1 Background

With the development of highly sensitive micro-methods of amino acid analysis and protein sequencing, electrophoretic procedures have become the method of choice for the preparation and purification of proteins for subsequent chemical characterization. This approach has an important advantage over conventional protein purification strategies; namely that starting with a complex protein mixture containing up to several thousand species, a particular protein can be isolated to purity in a single step using either one- or two-dimensional electrophoresis.

Diffusion or electroelution (see Section 12.2) can be used to recover proteins from gels for subsequent chemical characterization, but each suffers from several limitations: (i) they are prolonged procedures, (ii) recovery can be variable, (iii) only a few samples can be processed simultaneously, (iv) proteins can be contaminated with detergents, salts and other gel constituents (e.g. Tris, glycine, acrylamide), (v) proteins can be degraded, and (vi) chemical modification of proteins can occur. These problems can be overcome using procedures based on the use of Western blotting (see Chapter 11) and this approach, which is more widely used, will be described in Section 12.3.

12.2 Recovery of proteins from gels

Although relatively small amounts of proteins are contained in individual zones after electrophoresis, if these proteins can be recovered there are many techniques which can be applied to their further characterization (e.g. amino acid compositional analysis, amino acid sequence analysis, peptide mapping, production of antibodies, determination of enzymic activity, estimation of biological activity). Two methods are available for the recovery of proteins from a gel and involve elution either by diffusion or by electroelution. The use of gels prepared with soluble cross-linking agents (see Section 3.6) is also a possibility, but the reagents required for gel solubilization can be damaging to proteins and are often incompatible with subsequent procedures.

12.2.1 Elution by diffusion

In this approach the gel is sliced, following electrophoresis, into small segments and macerated by chopping with a fine scalpel or by homogenization. The resulting gel slurry is then eluted, usually overnight at 4°C, with an appropriate buffer. The volume of added buffer must generally be kept small in order to minimize dilution of the protein, but this obviously limits the recovery. It is a good idea to ensure mixing during the elution process, for example, by attaching the tubes containing the samples to a vertical rotating disk. The eluted proteins are then recovered from the gel debris by centrifugation. Elution efficiency can be improved using elevated temperatures, providing the protein is stable under these conditions. The efficiency of extraction can also be improved by adding urea (4–8 M) or 0.1% (w/v) SDS to the elution buffer, but only if the resulting protein denaturation is compatible with the techniques to be used subsequently. For certain procedures, SDS and other additives must be removed after protein elution, and a variety of methods have been described to achieve this [1]. The simplest approach is to precipitate the eluted protein with 10–15% (w/v) TCA and to remove residual SDS from the precipitate by washing with acetone or ethanol.

12.2.2 Electrophoretic elution

In this approach, proteins are eluted out of the gel slices into a chamber which makes electrical contact with a reservoir buffer via a dialysis membrane. In the simplest system the gel slices containing the protein of interest are mixed with a small amount of polyacrylamide or agarose gel and cast into a glass tube (such as that used for rod gel electrophoresis). A small sack made from dialysis tubing and containing electrophoresis buffer is then fastened with elastic bands over the end of the tube. The tube is then inserted into a tube gel electrophoresis apparatus and the protein eluted into the dialysis sack, from which it can subsequently be recovered. This method is generally more efficient if 0.1% (w/v) SDS is added to the elution buffer. Rather more sophisticated variants of this approach have been described [1,2], and commercial equipment is now available which allows several different gel segments to be eluted simultaneously into a series of elution chambers. One of these commercial designs is shown in *Figure 12.1*.

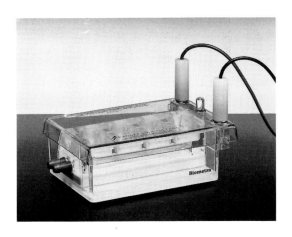

FIGURE 12.1: Commercial apparatus for electroelution of proteins from polyacrylamide gels. Photograph courtesy of Biometra.

12.3 Protein characterization on Western blots

12.3.1 Amino acid analysis

Amino acid analysis is a routine approach for determining the amino acid composition of a protein. Moreover, it is the best method for determining the absolute concentration of a protein. The sensitivity of the method has increased over the last 20 years from the 1 nmol level using the ninhydrin detection system, to the 50 pmol level with the phenylisothiocyanate (PITC) method, and to the 5–10 pmol level using the o-phthalaldehyde (OPA) or fluorescamine methods. The most sensitive methods currently available (as low as 500 fmol) use reverse-phase high-performance liquid chromatography (RP-HPLC) with pre-column derivitization with reagents such as OPA, dansyl chloride, dabsyl chloride, fluorenmethyl chloroformate, and PITC.

Gel electrophoresis, and in particular SDS–PAGE, combined with Western blotting has become the method of choice for purifying proteins and peptides for amino acid analysis. Some care must be exercised in the choice of blotting transfer buffer as the use of glycine and amino buffers such as Tris should be avoided to minimize background interference. The transfer buffer which is currently recommended is 10 mM 3-(cyclohexylamino)-1-propanesulfonic acid (CAPS), pH 11.0, containing 10% (v/v) methanol. The choice of blotting matrix is also critical as nitrocellulose is not compatible with the organic solvents used for amino acid analysis; elution of the bound proteins from nitrocellulose is necessary prior to analysis. It is, therefore, better to use polyvinylidene difluoride (PVDF) membranes which are compatible with organic solvents and which allow direct amino acid analysis of proteins while still bound to the membrane [3,4].

After electroblotting, the protein band or spot of interest is simply excised, and the protein on its PVDF support is subjected to acid hydrolysis. After hydrolysis is complete, the amino acids are extracted from the membrane, dried and submitted to amino acid analysis by a suitably sensitive method (e.g. RP-HPLC with OPA detection) [3,4]. It is also recommended to submit a piece of the same blotting membrane, but having no bound protein, to the same procedure as a control to account for contamination with glycine from the SDS–PAGE step.

12.3.2 N-terminal protein sequence analysis

Almost all protein sequence determination uses the method, known as Edman degradation, devised by Edman in 1949 [5] (*Figure 12.2*). However, protein sequence analysis did not become a practicable routine procedure until two decades later with the development of automated sequencing equipment [6]. These early machines required large amounts (100–1000 nmol) of proteins. Since then, progress in sequenator technology and the optimization of the degradation chemistry has resulted in the development of the current generation of gas–liquid phase and solid-phase sequenators which are capable of determining limited N-terminal sequences from as little as 10–20 pmol of a highly purified protein at a speed of approximately 1 amino acid per 30–50 min (see *Table 12.1*). This amount corresponds to 0.5–1 µg of pure protein of $M_r = 50 \times 10^3$ molecular mass. Protein zones following gel electrophoresis often contain proteins in these amounts, so that techniques of one- and two-dimensional electrophoresis, in conjunction with Western blotting, have become the methods of choice for the preparation and purification of proteins for sequence analysis. This method will be described briefly here, but for a full account the reader is referred to two recent reviews [2,7].

Most gel electrophoresis techniques can be used as a preparative procedure for protein sequence analysis, but SDS–PAGE and 2-D PAGE are the methods which have most often been used. The amount of sample applied is of great importance, particularly if complex mixtures are to

TABLE 12.1: The evolution of protein sequencing sensitivity

Protein required (nmol)	Mass (µg*)	Instrument
100–1000	5000–50 000	Spinning Cup (1967)
10	500	Commercial Spinning Cup (1971)
0.05–0.1	2.5–5	Modified Spinning Cup (1978–1980)
0.01–0.02	0.5–1	Gas–liquid phase (1980)
0.005	0.25	Optimized gas–liquid phase (1990)
0.005	0.25	Optimized solid phase (1990)
0.0005	0.025	Equipment in development

*Mass of a $M_r = 50 \times 10^3$ protein, assuming 100% sequenceable material.
Modified from reference [2].

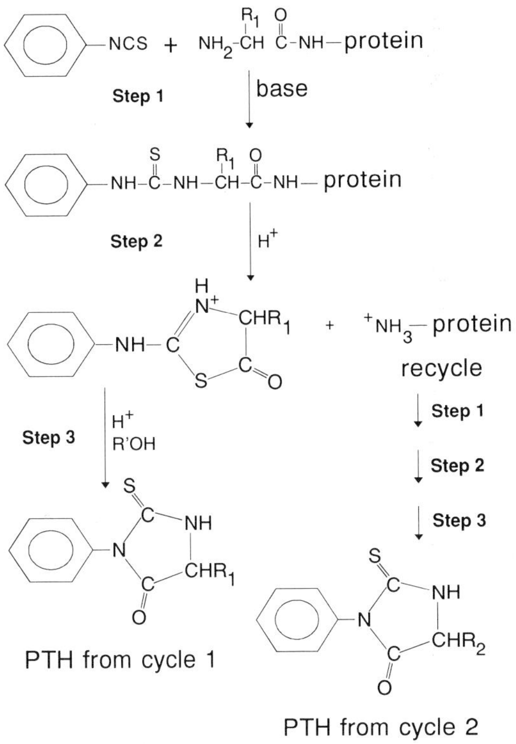

FIGURE 12.2: *Edman chemistry for the stepwise degradation of proteins from the N-terminus. In the first step, the α-amino group of the polypeptide chain is reacted with phenylisothiocyanate (PITC). The resulting PTC–protein is then cleaved under acidic conditions (step 2), to release a derivative of the N-terminal amino acid from the polypeptide chain. This process of coupling and cleavage is repeated on the newly revealed N-terminus of the shortened polypeptide chain. The amino acid derivatives released at each cycle of degradation are converted to stable phenylthiohydantoin (PTH) derivatives (step 3) and analyzed by RP-HPLC. In this way, the identity of the amino acid at each position of the peptide chain is determined. Redrawn from reference [2] with permission from VCH.*

be analyzed. In general, a load suitable for use with a general protein stain such as Coomassie brilliant blue is required, which should be of the order of several hundred micrograms. Such high protein loads can be particularly problematical for 2-D PAGE separations due to loss of resolution using standard equipment for analytical 2-D PAGE. One approach to this problem is to use glass tubes of a greater (e.g. 3 mm) internal diameter for the first dimension IEF in conjunction with correspondingly thicker second dimension SDS–PAGE gels [8]. The higher protein loading capacity of IPG IEF gels can be used to advantage here: for example, Hanash and colleagues [9] have applied up to 1 mg of cellular proteins to pH 4–7 IPG IEF gel strips without loss of resolution in the final 2-D maps.

Both buffer-tank and semi-dry electroblotting procedures are compatible with subsequent protein sequencing, but the choice of transfer buffer and blotting matrix is critical. The use of transfer buffers containing glycine (or other amino acids) should be avoided. The two buffer systems currently recommended are (i) 50 mM Tris, 50 mM borate, pH 8.5, containing 20% (v/v) methanol, or (ii) 10 mM CAPS, pH 11.0, containing 10% (v/v) methanol.

Nitrocellulose membranes are not compatible with the reagents and organic solvents used in automated protein sequencing, which has resulted in the development of a range of alternative matrices (*Table 12.2*). Two characteristics of the blotting matrix must be considered. Firstly, it is important that the membrane has a good protein binding capacity, and secondly the matrix must perform well during automated sequence analysis. Unfortunately, these two properties do not necessarily go together [10]. In automated sequencers using gas–liquid phase technology, pure protein samples are applied directly to glass fiber (GF/C) filter disks, treated with polybrene, resulting in theoretical initial yields of 50–70% and average repetitive yields of 90–95%. These membranes have pore properties which allow good solvent flow during gas–liquid phase sequencing and optimal reaction with the protein to be sequenced. However, these filters cannot be used for electroblotting. This prompted the development of derivatized glass fiber filters, which could be used for electroblotting while still retaining good solvent flow properties (*Table 12.2*). These filters still do not have the high protein binding capacity for Western blotting [10]; some of them must be activated prior to use and their inherent fragility makes handling difficult (particularly in the case of large format 2-D gels). Consequently, more robust supports, particularly those based on the use of PVDF (see Section 11.3), are recommended [11,12] (*Table 12.2*). However, the small pore size of PVDF membranes can cause problems of restricted solvent flow using the standard 'flow-through'

TABLE 12.2: *Electroblotting matrices for the preparation of proteins for sequence analysis*

Matrix	Description	Supplier
QA-GF	Glass fiber covalently modified with	—
AP-GF	quaternary (QA) or aliphatic (AP) groups	
PCGM-1	Glass fiber non-covalently modified with polybases	Janssen Life Science
Glassybond	Siliconized glass fiber	Biometra
Immobilon P	Polyvinylidene difluoride	Millipore
Fluorotrans	Polyvinylidene difluoride	Pall BioSupport
ProBlott	Polyvinylidene difluoride	Applied Biosystems
Westrans	Polyvinylidene difluoride	Schleicher & Schuell
PP-20	Polypropylene	Schleicher & Schuell
DITC-GF	Glass fiber covalently modified with *p*-phenylene-diisothiocyanate	—

Addresses for suppliers can be found in Appendix B.

reaction cartridge of gas–liquid phase sequenators. This problem, which does not occur in solid phase sequenators [2], has been solved by a modified cartridge design which permits a vertical cross-flow reagent path [13].

After the Western blotting step is complete, the proteins to be sequenced must be detected. The PVDF and polypropylene based matrices can be stained with Coomassie brilliant blue or Amido black and, provided that excess dye is removed from the membrane, the dye which remains bound to the protein does not interfere with the sequencing process. Stains such as Fast green and Ponceau can also be used; although not very sensitive they have the advantage that they can be easily removed from proteins after detection. These staining methods cannot be used with matrices based on glass fiber, so that fluorescent methods involving 3,3′-dipentyloxacarbocyanine iodide or fluorescamine must be used [2,7].

After detection, the protein band or spot of interest is simply excised and introduced directly into the reaction cartridge of the protein sequenator to obtain the corresponding N-terminal amino acid sequence for that protein. The protein can then be identified, or homologous proteins suggested, by comparative sequence analysis using the international protein and DNA sequence databases. If the particular protein of interest has not been previously characterized, the availability of partial amino sequence information allows several powerful approaches to the further characterization of that protein by techniques of protein chemistry or molecular biology (*Figure 12.3*). Indeed, this approach can be seen as providing an interface between studies of gene expression at the protein level by techniques such as 2-D PAGE and computer databases (see Section 10.4) with the rapidly expanding body of information concerning the complexity of gene expression at the DNA level [14].

12.3.3 Internal protein sequence analysis

Generally, protein samples prepared by electroblotting can be sequenced at high efficiencies. In addition to the high repetitive stepwise yields, sequencing runs of blotted proteins yield PTH analyses with very low backgrounds of by-product peaks. In practice, however, many proteins are found to yield no N-terminal sequence information because they lack a free α-amino group (i.e. the N-terminus is 'blocked'). Modifications resulting in blockage of the N-terminus frequently occur post- or co-translationally within the cell. Perhaps as many as 50–75% of all cellular proteins are blocked and these modifications, which are of physiological importance, commonly involve the addition of formyl, acetyl or acyl groups or cyclization to pyrrolidone carboxylic acid. N-terminal blockage can also result as an artifact during sample work-up prior to sequencing, but this problem can be minimized by the use of high purity, quality controlled reagents and by avoiding conditions known to be amino reactive. *In vivo* N-terminal blockage is much more difficult to deal with. Specific chemical and enzymic methods exist for unblocking, but cannot be applied to samples prepared by electrophoresis due to the low amount of protein available and the lack of knowledge of the blockage group involved.

The best approach to the generation of protein sequence information from N-terminally blocked proteins is to use chemical (e.g. cyanogen bromide, BNPS-skatole), or enzymatic (e.g. trypsin, chymotrypsin, subtilisin, papain) methods to cleave the intact polypeptide chain to generate shorter peptides which can be isolated and sequenced. The availability of internal sequence information has additional advantages for sequence homology searches and gene cloning strategies [2]. Proteins separated by electrophoresis can be recovered prior to cleavage by electroelution from the gel (see Section 12.2.2), by elution from a Western blot (e.g. from PVDF with acetonitrile) [2], or by liquid pressure extraction [2]. However, most of these procedures suffer from problems of contamination and poor recoveries. This has resulted in the development of two alternative strategies.

The first approach combines the method of Cleveland [15] for proteolytic cleavage during SDS–PAGE of a gel slice containing the protein of interest (see Section 6.8.2) with electroblotting of the

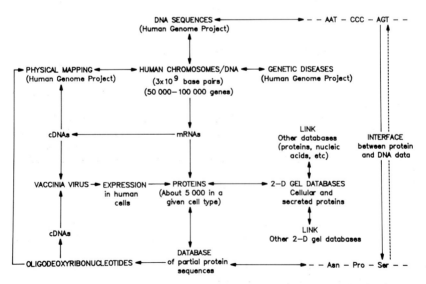

FIGURE 12.3: *Interface between comprehensive 2-D gel protein databases and the human genome mapping and sequencing program. Reproduced from reference [15] with permission from VCH.*

resulting separated peptides for subsequent sequence analysis [16]. This method is only suitable for relatively large peptides ($M_r > 5000$) due to the limited ability of SDS–PAGE to resolve small peptides and their poor binding capacity to blotting matrices. An alternative approach suitable for small peptides generated by exhaustive cleavage of protein in gel slices is to extract the peptides with trifluoroacetic acid (TFA) for separation by RP-HPLC [17].

The second approach is the enzymatic or chemical fragmentation of proteins *in situ* on blotting membranes (nitrocellulose or PVDF) after electrotransfer [18]. Cleavage fragments are then released using TFA and the peptides separated by narrow- or micro-bore RP-HPLC [2]. Selected peptides can then be collected directly onto standard polybrene-treated glass fiber filter disks for direct application to the protein sequenator. This procedure is highly efficient and the determination of multiple internal stretches of sequence requires only two to three times more protein than does N-terminal protein sequence analysis.

References

1. Andrews, A.T. (1986) *Electrophoresis: Theory, Techniques and Biochemical and Clinical Applications*. Clarendon Press, Oxford.
2. Aebersold, R. (1991) in *Advances in Electrophoresis* (A. Chrambach, M.J. Dunn and B.J. Radola, eds). VCH, Weinheim, Vol. 4, pp. 81–168.
3. Tous, G.I., Fausnaugh, J.L., Akinyosoye, O., Lackland, H., Winter-Cash, P., Vitorica, F.J. and Stein, S. (1989) *Anal. Biochem.*, **179**, 50–55.
4. Nakagawa, S. and Fukuda, T. (1989) *Anal. Biochem.*, **181**, 75–78.
5. Edman, P. (1949) *Arch. Biochem. Biophys.*, **22**, 475–483.
6. Edman, P. and Begg, G. (1967) *Eur. J. Biochem.*, **1**, 80–91.
7. Simpson, R.J., Moritz, R.L., Begg, G.S., Rubira, M.R. and Nice, E.C. (1989) *Anal. Biochem.*, **177**, 221–236.
8. Baker, C.S., Corbett, J.M., May, A.J., Yacoub, M.H. and Dunn, M.J. (1992) *Electrophoresis*, **13**, 723–726.
9. Hanash, S.M., Strahler, J.R., Neel, J.V., Hailat, N., Melhem, R., Keim, D., Zhu, X.X., Wagner, D., Gage, D.A. and Watson, J.T. (1991) *Proc. Natl Acad. Sci. USA*, **88**, 5709–5713.
10. Baker, C.S., Dunn, M.J. and Yacoub, M.H. (1991) *Electrophoresis*, **12**, 342–348.
11. Pluskal, M.G., Przekop, B., Kavonian, M.R., Vecoli, C. and Hicks, D.A. (1986) *BioTechniques*, **4**, 272–283.
12. Matsudaira, P. (1987) *J. Biol. Chem.*, **262**, 10035–10038.
13. Sheer, D.G., Yuen, S., Wong, J., Wasson, J. and Yuan, P.M. (1992) *BioTechniques*, **11**, 526–533.
14. Celis, J.E., Rasmussen, H.H., Madsen, P., *et al.*, (1992) *Electrophoresis*, **13**, 893–959.
15. Cleveland, D.W., Fischer, S.G., Kirschner, M.W. and Laemmli, U.K. (1977) *J. Biol. Chem.*, **252**, 1102–1106.
16. Kennedy, T.E., Gawinowicz, M.A., Barzilai, A., Kandel, E.R. and Sweatt, J.D. (1988) *Proc. Natl Acad. Sci. USA*, **85**, 7008–7012.
17. Eckerskorn, C. and Lottspeich, F. (1989) *Chromatographia*, **28**, 92–94.
18. Aebersold, R., Leavitt, J., Saavedra, R., Hood, L.E. and Kent, S.B.H. (1987) *Proc. Natl Acad. Sci. USA*, **84**, 6970–6974.

Appendix A
Glossary

Ampholytes: molecules, including proteins, which are negatively charged in solutions where the pH is above their isoelectric point (pI) and positively charged in solutions where the pH is below their pI.

Anolyte: the electrolyte solution used at the anode.

Blotting: the transfer of macromolecules from a gel in which they have been separated to the surface of an inert support membrane. The technique was originally devized by Professor E.M. Southern for the transfer of DNA, so this technique became known as Southern blotting. On this basis, techniques which were subsequently developed for the transfer of RNA and proteins became known as Northern and Western blotting, respectively.

C (per cent): proportion of cross-linking agent (e.g. Bis) in total gel monomer mixture (acrylamide + Bis) in grams per 100 g of mixture.

Carbamylation: interaction of amino groups of proteins with cyanate ions formed by the breakdown of urea.

Carrier ampholytes: a mixture of low molecular weight amphoteric substances used to establish the pH gradient in IEF.

Catholyte: the electrolyte solution used at the cathode.

Co-electrophoresis: the simultaneous electrophoretic separation of a mixture of two different protein samples.

Electroblotting: the electrophoretic transfer of macromolecules from a gel in which they have been separated to the surface of an inert support membrane.

Electroendosmosis (EEO): EEO is defined as the flow of an ionic solvent adjacent to a fixed, charged surface resulting from the application of an electric field. All electrophoretic support media (e.g. paper, agarose, polyacrylamide, glass) possess fixed charged groups which are ionized at the pH of the buffers normally used for electrophoresis. These groups are usually negatively charged and, therefore, attracted towards the anode. However, these charged groups are immobile as they are fixed on the support matrix. This is compensated for by a migration of H^+ ions (as hydrated protons, H_3O^+) towards the cathode. This effect results in a flow of solvent relative to the support medium. EEO is usually detrimental to electrophoretic separation, but can be exploited to advantage in certain techniques (e.g. immunoelectrophoresis, capillary electrophoresis).

Fluor: a molecule which will emit photons when irradiated by an appropriate form of energy (e.g. β-particles, X-rays, light).

Histones: small basic proteins which bind tightly to eukaryotic DNA. A striking feature of histones is their high content of positively charged side chains; about one in four residues is either lysine or arginine. Five major species exist (H1, H2A, H2B, H3, H4), but each type can exist in a variety of forms due to post-translational modifications of certain side chains.

HPLC: high-performance liquid chromatography. A set of chromatography techniques and equipment designed and optimized for the isolation and purification of small quantities of biomolecules.

Immunoelectrophoresis: a procedure in which proteins and other antigenic substances are characterized by both their electrophoretic migration in a gel (usually agarose) and their immunological properties. There are many variations of technique but they are all based on the electrophoretic migration of antigens in an antibody-containing gel and specific immunoprecipitation of the antigens by means of corresponding precipitating antibodies.

Immunoprecipitate: insoluble complex formed between an antigen and a corresponding precipitating antibody.

Isoelectric point (pI): the pH at which the net charge on an amphoteric macromolecule is zero.

Joule heating (H): heat generated by the passage of electric current; dependent on both voltage (V, in volts) and current (I, in amperes) with the relationship, $H = V \times I$ (J/sec).

Lectins: animal or plant proteins or glycoproteins of non-immune origin which bind specifically with high affinity to carbohydrate residues.

Ligand: any substance which binds specifically to a receptor.

Molecular mass: the mass of one molecule of a substance expressed in daltons (Da) or atomic mass units. The dalton is defined as one-twelfth of the mass of one atom of ^{12}C.

Non-histone nuclear proteins: proteins associated with eukaryotic DNA which are not histones. They are present only at about 5% of the level of the histones.

Non-restrictive gel: gel with a large effective pore size (i.e. low %T at a fixed %C; or very high %C at a fixed %T) so that there is minimum physical restriction of the passage of a macromolecule through the gel matrix.

Pre-electrophoresis: electrophoresis of the separation gel alone, prior to the application of the sample, to remove any contaminants (e.g. unreacted polymerization agents).

Pre-focusing: optional step in IEF in which the gel is subjected to electrophoresis, prior to the application of sample, to allow establishment of the pH gradient.

Relative mobility (R_f): the mobility of the protein of interest measured with reference to a marker protein or tracking dye: R_f = distance migrated by protein/distance migrated by dye. The value of R_f is, therefore, always equal to or less than unity.

Relative molecular mass (M_r): the ratio of the mass of a molecule to one-twelfth of the mass of the nuclide ^{12}C. It is, therefore, dimensionless.

Reporter molecule: molecule covalently linked to a ligand which can be used to generate a signal for the detection of the binding of the ligand to its receptor.

T (%): total concentration of monomer (acrylamide + Bis), in grams per 100 ml, used in polyacrylamide gel preparation.

Zwitterions: macromolecules (e.g. proteins, some detergents) which possess both anionic and cationic groupings as part of their structure.

Appendix B

Suppliers

1. Chemicals, reagents and suppliers

Aldrich Chemical Co. Inc., 1001 W. St Paul Ave., PO Box 355, Milwaukee, WI 53201, USA.

Amersham International, Amersham Plc., Little Chalfont, Amersham, Buckinghamshire HP7 9NA, UK.

Bio-Rad Laboratories, 1414 Harbor Way South, Richmond, CA 93804, USA.

Boehringer Mannheim GmbH, Sandhoferstrasse 116, PO Box 310120, D-6800 Mannheim 31, Germany.

Calbiochem Corp., 10933 Torrey Pines Rd., La Jolla, CA 92037, USA.

Daiichi Pure Chemical Co. Ltd, 13-5, Nihombashi 3-chome, Chuoku, Tokyo 103, Japan.

Eastman Kodak Co., 343 State St., B-701, Rochester, NY 14652-3512, USA.

Fluka Chemie AG, Indistriestrasse 25, CH-9470 Buchs, Switzerland.

FMC BioProducts, 191 Thomaston St., Rockland, ME 04841, USA.

Gibco BRL, PO Box 9418, Gaithersburg, MD 20898, USA.

ICN Biomedicals Inc., 3300 Hyland Ave., Costa Mesa, CA 92626, USA.

Merck Ltd, Broom Rd, Poole, Dorset BH12 4NN, UK.

Millipore Corp., 80 Ashby Road, Bedford, MA 01803, USA.

NBS Biologicals, Ediston House, 163 Dixons Hill Rd, North Myms, Hatfield, Hertfordshire AL9 7JE, UK.

Pall BioSupport Corp., 77 Crescent Beach Rd, Glen Cove, NY 11542, USA.

Pharmacia LKB Biotechnology AB, Bjorkgatan 30, S-75182 Uppsala, Sweden.

Polysciences Inc., 400 Valley Road, Warrington, PA 18976, USA.

Sartorius AG, PO Box 3243, D-3400 Göttingen, Germany.

Schleicher and Schuell GmbH, PO Box 4, D-3354 Dassel, Germany.

Serva Feinbiochemica GmbH, Carl-Benz-Strasse 7, D-6900 Heidelberg, Germany.

Sigma Chemical Co., 3050 Spruce St., St Louis, MO 63178, USA.

2. Equipment

ATTO Corp., 2-3, Hongo 7-chome, Bunkyo-ku, Tokyo 113, Japan.

Biometra Biomedizinische Analytik GmbH, Rudolf-Wessel-Strasse 30, D-3400 Göttingen, Germany.

Bio-Rad Laboratories, 1414 Harbor Way South, Richmond, CA 93804, USA.

DESAGA GmbH, PO Box 101969, D-6900 Heidelberg, Germany.

E-C Apparatus Corp., 3831 Tyrone Blvd N., St Petersburg, FL 33709, USA.

Fisher Scientific, 711 Forbes Ave., Pittsburgh, PA 15219, USA.

Genetic Research Instrumentation Ltd, Gene House, Dunmow Rd, Felsted, Dunmow, Essex CM6 3LD, UK.

Hoefer Scientific Instruments, 654 Minnesota St., San Francisco, CA 94107-3027, USA.

Millipore Corp., 80 Ashby Road, Bedford, MA 01803, USA.

Pharmacia LKB Biotechnology AB, Bjorkgatan 30, S-75182 Uppsala, Sweden.

Shandon Scientific Ltd, Chadwick Rd, Astmoor, Runcorn, Cheshire WA7 1PR, UK.

3. Software package for generation of immobilized pH gradients

Fluka Chemie AG, Indistriestrasse 25, CH-9470 Buchs, Switzerland.

4. Densitometers

Beckman Instruments Inc., 2500 Harbor Blvd, Box 3100, Fullerton, CA 92634-3100, USA.

Biometra Biomedizinische Analytik GmbH, Rudolf-Wessel-Strasse 30, D-3400 Göttingen, Germany.

Bio-Rad Laboratories, 1414 Harbor Way South, Richmond, CA 93804, USA.

CAMAG, Sonnenmattstrasse 11, CH-4132 Muttenz, Switzerland.

DESAGA GmbH, PO Box 101969, D-6900 Heidelberg, Germany.

Fuji Photo Film Co. Ltd, 26-30, Nishiazabu 2-chome, Minato-ku, Tokyo 106, Japan.

Helena Laboratories, PO Box 752, Beaumont, TX 77704-0752, USA.

Hoefer Scientific Instruments, 654 Minnesota St., San Francisco, CA 94107-3027, USA.

Molecular Dynamics, 880 E. Argues Ave., Sunnyvale, CA 94025, USA.

Pharmacia LKB Biotechnology AB, Bjorkgatan 30, S-75182 Uppsala, Sweden.

5. Software packages for analysis of 2-D gels

Applied Imaging, Hylton Park, Wessington Way, Sunderland, Tyne and Wear SR5 3HD, UK.

Biolog, 100 Rue des Artisans, Z.A. de Buc, F-78530-Buc, France.

Biometra Ltd, PO Box 167, Maidstone, Kent ME14 2AT, UK.

Indiana Biotech, Highland, IN, USA.

Large Scale Biology Corporation, 9620 Medical Center Drive, Rockville, MD 20850, USA.

Millipore Corporation, 80 Ashby Road, Bedford, MA 01803, USA.

Protein and DNA Imageware Systems, 405 Oakwood Road, Huntington Station, NY 11746, USA.

Quanti-Gel Corporation, 1202 Ann Street, Madison, WI 53713, USA.

Technology Resources Inc., Nashville, TN, USA.

Appendix C
Further reading

1. Books

✻ Allen, R.C., Saravis, C.A. and Maurer, H.R. (1984) *Gel Electrophoresis and Isoelectric Focusing of Proteins*. Walter de Gruyter, Berlin.

✻Andrews, A.T. (1986) *Electrophoresis: Theory, Techniques and Biochemical and Clinical Applications*. Clarendon Press, Oxford.

Baldo, B.A. and Tovey, E.R. (1989) (eds) *Protein Blotting. Methodology, Research and Diagnostic Applications*. Karger, Basel.

✻Celis, J.E. and Bravo, R. (1984) *Two-Dimensional Gel Electrophoresis of Proteins*. Academic Press, Orlando, FL.

Chrambach, A. (1985) *The Practice of Quantitative Gel Electrophoresis*. VCH, Weinheim.

Dunbar, B.S. (1987) *Two-Dimensional Electrophoresis and Immunological Techniques*. Plenum Press, New York.

Dunn, M.J. (1986) (ed.) *Gel Electrophoresis of Proteins*. Wright, Bristol.

Hames, B.D. and Rickwood, D. (1990) (eds) *Gel Electrophoresis of Proteins: a Practical Approach*. IRL Press, Oxford.

Mosher, R.A., Saville, D.A. and Thormann, W. (1992) *The Dynamics of Electrophoresis*. VCH, Weinheim.

Righetti, P.G. (1983) *Isoelectric Focusing: Theory, Methodology and Applications*. Elsevier, Amsterdam.

✻Righetti, P.G. (1990) *Immobilized pH Gradients: Theory and Methodology*. Elsevier, Amsterdam.

2. Journals and Annuals

Advances in Electrophoresis, Chrambach, A., Dunn, M.J. and Radola, B.J. (eds). VCH, Weinheim.

Analytical Biochemistry. Academic Press, San Diego, CA.

Applied and Theoretical Electrophoresis. Macmillan Press, London.

Electrophoresis. VCH, Weinheim.

Journal of Biochemical and Biophysical Methods. Elsevier, Amsterdam.

Journal of Chromatography. Elsevier, Amsterdam.

Index

Agarose gel, 10, 25
Alcian blue stain, 124
Amido black stain, 112, 146
Amidosulfobetaine detergents, 48, 93
Amino acid analysis, 155–156
Amino acid structure, 3
Ammonium persulfate, 15
AMPPD (disodium 3-(4-methoxyspiro
 [1,2-dioxetane-3-2′-tricyclo-
 [3.3.3.1^{3,7}]decan]4-yl)phenyl
 phosphate), 150–153
Antibody,
 use for antigen detection in gels, 139
 use for antigen detection on
  Western blots, 139, 148–152
ANS (1-aniline-8-naphthalene
 sulfonate), 115
Autoradiography,
 direct, 121
 intensifying screens, 122

BAC (N,N′-bis-acrylylcystamine), 17
Basic proteins,
 analysis by 2-D PAGE, 94, 97
Bis (N,N′-methylene-bis-acrylamide),
 13
Blotting – see Western blotting
Buffer systems,
 choice of ionic strength, 32–33
 choice of pH, 31–32
 continuous, 8, 31–33, 52
 discontinuous (multiphasic), 8, 33–
  36, 52–53
 Western blotting, 145–146, 155

%C definition, 16
Carbamylation of proteins, 73, 81, 100
Cathodic drift, 74, 75, 94

Chemical characterization of proteins
 after electrophoresis,
 amino acid analysis, 155–156
 background, 153
 internal protein sequence analysis,
  cleavage methods,
   in gel, 160
   on blot, 161
  N-terminal blockage, 159–160
  Western blotting, 161
 N-terminal protein sequence
  analysis,
  blotting membranes, 145, 158–159
  choice of electrophoretic method,
   156
  N-terminal blockage, 159–160
  sensitivity, 156
 protein recovery
  electrophoretic elution, 154
  elution by diffusion, 154
Chemiluminescent detection of
 immunoblots, 150–153
CHAPS (3-[(cholamidopropyl)
 dimethylammonio]-1-propane
 sulfonate), 47, 67, 93
Charge density, 9
Colloidal gold stain, 146
Colloidal iron stain, 146
Coomassie brilliant blue G-250, 114
Coomassie brilliant blue R-250, 113, 146
Continuous zone electrophoresis, 31–33
Cross-linking agents, 13, 17, 124
CTAB (cetyltrimethylammonium
 bromide), 48

Dansyl hydrazine stain for
 glycoproteins, 124
DATD (N,N′-diallyltartardiamide), 17,
 124

Densitometry – see Quantitative analysis
Detection methods,
  Amido black, 112
  Coomassie brilliant blue G-250, 114
  Coomassie brilliant blue R-250, 113
  enzymes, 125–126
  fixation, 111–112
  fluorescent stains, 115
  glycoproteins, 124
  lipoproteins, 125
  phosphoproteins, 125
  pre-labeling with fluorophores, 115
  radioactive proteins
    autoradiography, 121
    dual isotope methods, 122–123
    electronic detection methods, 123, 132–133
    fluorography, 121–122
    gel drying, 120
    gel slicing and counting, 123–124
    intensification screens, 122
    phosphor-imaging systems, 133–135
  silver staining,
    advantages and problems, 115–116
    color methods, 118
    enhancement, 117
    fixation, 116–117
    mechanism, 118–119
    glycoproteins, 124
    procedures, 117–118
  Western blotting, 139–152
  see also Western blotting
DHEBA (N,N'-(1,2-dihydroxyethylene) bis-acrylamide), 17, 124
Detergents,
  anionic, 48
  cationic, 48–49
  choice of, 43–45
  non-ionic, 45, 67
  types of, 43
  zwitterionic, 47–48, 67
Dissociation constant (K_d), 82–83
Disulfide bond reduction, 41

EDIA (ethylenediacrylate), 17
Electroblotting – see Western blotting
Electroelution, 154
Electroendosmosis, 10, 74, 75, 94
Electronic detection methods, 123, 132–135

Electrophoretic mobility,
  definition, 7
  effect of ionic strength, 8
  effect of pH, 8
Enzyme detection, 125–126

Fast green FC, 146
Ferguson plot, 36–37, 38
Fixation, 111–112, 116–117
Fluorescamine, 115
Fluorography, 121–122

GelBond film, 23, 69
Gel drying, 120
Gel recovery, 111
Gene expression, 2
Genetic mutation analysis, 81–82
Glycoproteins
  detection in gels, 124
  detection on blots, 139, 148
  molecular mass by SDS–PAGE, 57–58
Gradient polyacrylamide gels,
  advantages, 26
  gradient making devices, 27–28
  ready-made gels, 29
  linear gradient gels, 27
  native protein separation, 37–38
  non-linear gradient gels, 30
  SDS–PAGE, 55–57
  size fractionation range, 27
  transverse, 38, 42–43

Histones,
  acid–urea gels, 42
  interaction with IPG matrices, 80
  SDS–PAGE, 58
  2-D PAGE, 106–107
Homogeneous polyacrylamide gels, 25

IEF – see isoelectric focusing
Immunoelectrophoresis, 25
Immunofixation, 139–140
India ink, 146
Ionic strength
  effect on Joule heating, 8
  effect on mobility, 8
Isoelectric focusing (IEF),
  carrier ampholyte pH gradients,
    apparatus, 23, 24, 69
    basic principles, 65–67
    carrier ampholytes 67–69, 74

Isoelectric focusing (IEF)—cont.
  cathodic drift, 74
  charge-balanced matrices, 74
  detergents, 67
  electrophoresis, 70
  estimation of pH gradients, 71–73, 100
  first dimension of 2-D PAGE, 93–95
  gel composition, 67
  gel concentration, 67
  gel preparation, 23, 69
  limitations of method, 73–74
  pI marker proteins, 72–73, 100
  ready-made gels, 69
  running conditions, 70–71
  sample application, 70
  urea, 70
  immobilized pH gradients (IPG),
    basic principles, 74–75
    cathodic drift, 75
    estimation of pH gradients, 81, 100
    extended pH gradients, 76
    first dimension of 2-D PAGE, 95–99
    gel preparation, 77–78
    hybrid (mixed-bed) IEF, 77, 95
    Immobiline reagents, 75, 77
    narrow and ultra-narrow pH gradients, 76
    non-linear pH gradients, 76
    problems, 77
    ready-made gels, 78
    reswelling dry gels, 79
    running conditions, 80–81
    sample application, 79–80
Isoelectric point (pI), 5, 65, 67, 72–73, 81
Isoenzymes, 10

Joule heating, 8, 33, 142

Lectins,
  detection of glycoproteins in gels, 124
  detection of glycoproteins on Western blots, 139, 148
Lipoprotein detection, 125
Luminol, 150

Macromolecular interactions,
  titration curve analysis, 82–83
MDPF (2-methoxy-2,4-diphenyl-3(2H)-furanone), 115

Membrane proteins,
  analysis by 2-D PAGE, 109
Miniaturized electrophoresis apparatus, 21
Molecular mass determination,
  denatured proteins (SDS–PAGE),
    anomalous behaviour, 57–58
    gradient gels, 55–57
    homogeneous gels, 54–55
    limitations, 57–58
    marker polypeptides, 54–55
    small polypeptides, 57
  native proteins,
    Ferguson plot, 36–37, 38
    homogeneous gels, 36–37
    pore gradient gels, 37–38
    transverse gradient gels, 38
Molecular sieving, 25
Multiphasic zone electrophoresis, 33–36

Native PAGE,
  apparatus,
    rod gels, 18–19
    slab gels, 19–23
  buffer systems, 8, 31, 33
  Ferguson plot, 36–37, 38
  gel concentration, 16, 26
  gradient gels, 26, 37–38
  homogeneous gels, 25, 36–37
  molecular mass determination, 36–38
  non-ionic detergents, 45
  sample application, 20, 24
  transverse gradient gels, 38, 42–43
  ultra-thin gels, 23
  urea, 41–43
  zwitterionic detergents, 47
Nitrocellulose membranes, 144–145, 155
Nonidet P-40, 45, 92
Northern blotting, 140

Octyl-β-D-glucopyranoside, 93
OPA (o-phthalaldehyde),
  amino acid analysis, 155
  protein detection, 115

Peptide separation by SDS–PAGE, 52, 62–63
Peptide mapping,
  need for, 59–60
  primary gel system, 61–62
  protein cleavage,

Peptide mapping—*cont.*
  chemical, 62
  enzymatic, 62
  secondary gel system, 62–63
  strategy, 60
Periodic acid–Schiff (PAS) stain for
    glycoproteins, 124
pH,
  effect on mobility, 8
  effect on PAGE separations, 31–32
PhastSystem, 25
Phosphoprotein detection, 125
Phosphor-imaging systems, 133–135
PITC (phenylisothiocyanate), 155
pK analysis by titration curves,
    83–84
Polyacrylamide gel
  advantages, 10
  cross-linking agents, 13, 17
  polymerization catalysts, 15–16
  polymerization temperature, 16
  pore size, 13, 16
  properties, 13
  purity, 15
  structure, 13, 16
  toxicity, 13
Polymerization catalysts, 15
Ponceau S, 146
Pore gradient (limit) electrophoresis,
    37–38
Pore size, 13, 16
Power supplies, 9, 70, 80, 104
Pre-electrophoresis, 15
Protease inhibitors, 89
Protein,
  charge properties, 5
  isoelectric point, 5
  primary structure, 3
  quaternary structure, 4
  recovery from gels, 153–154
  secondary structure, 3
  sequence analysis
    internal, 159–161
    N-terminal, 156–159
  tertiary structure, 4
PVDF (polyvinylidene difluoride)
    membranes, 145, 155, 158

Quantitative analysis,
  gel imaging,
    CCD array scanners, 131–132

Quantitative analysis—*cont.*
    densitometers, 129–130
    laser densitometers, 130
    phosphor-imaging systems, 133–135
    radioisotope imagers, 132–133
    TV cameras, 130–131
  one-dimensional gels, 135–136
  two-dimensional gels, 136–138
  Western blots, 152

Radiolabelling of proteins, 119–120
Relative mobility (R_f), 36, 54–55
Retardation coefficient, 36
Riboflavin, 15, 49
Ribosomal proteins,
  analysis by 2-D PAGE, 105–106

SDS (sodium dodecyl sulfate),
  binding to proteins, 51, 57
  protein dissociation, 48, 51
  purity, 51
  sample preparation for 2-D PAGE, 92
SDS–PAGE,
  basic principles, 51
  choice of gel concentration, 51–52
  continuous buffer systems, 52
  discontinuous buffer systems, 52–53
  molecular mass determination,
    anomalous behaviour, 57–58
    gradient gels, 55–57
    homogeneous gels, 54–55
    limitations, 57–58
    marker polypeptides, 54–55
    small polypeptides, 57
  non-reducing conditions, 58–59
  peptides, 52
  peptide mapping, 59–63
  sample preparation, 53–54
  second dimension of 2-D PAGE, 102
  stacking, 52–53
  two-dimensional, 59
Silanization, 23
Silver stain – *see* Detection methods,
    silver stain
Slab gels,
  advantages, 20
  horizontal, 21
  miniaturized, 21
  preparation,
    homogeneous gels, 20, 25
    linear gradient gels, 27

Slab gels—*cont.*
  non-linear gradient gels, 30
  vertical, 21
Sodium deoxycholate, 48
Solubilization of polyacrylamide gels, 124
Southern blotting, 140
Stacking,
  mechanism, 34–35
  SDS–PAGE, 52–53
Staining methods – *see* Detection methods
Starch gels, 10

%T definition, 16
TEMED (N,N,N',N'-tetramethylethylenediamine), 15
Titration curves,
  analysis of macromolecular interactions, 82–83
  analysis of genetic mutations, 81–82
  basic principles, 81
  estimation of pK values, 83–84
Transverse gradient gel electrophoresis, 38, 42–43
Triton X-100, 45, 67
Tween 20, 148
Two-dimensional polyacrylamide gel electrophoresis (2-D PAGE),
  apparatus, 19, 103–104
  background, 87–88
  basic proteins, 94
  equilibration between dimensions, 100–101
  first dimension (IEF),
    estimation of pH gradient, 100
    isoelectric focusing using carrier ampholytes, 93–95
    isoelectric focusing using immobilized pH gradients, 95–99, 157
    gel recovery, 99–100, 111
    non-equilibrium pH gradient gel electrophoresis (NEPHGE), 94–95
    pI markers, 100
    preparation of rod (tube) gels, 93–94
    running conditions, 94, 99
    sample application, 94, 99
  gel size, 103
  histones, 106
  history, 88
  membrane proteins, 109
  native conditions, 104

Two-dimensional polyacrylamide gel electrophoresis (2-D PAGE)—*cont.*
  nuclear proteins, 106–108
  protein sequence analysis, 157, 159
  quantitative analysis, 136–138
  requirement for, 87
  resolution, 87–88, 103
  ribosomal proteins, 105–106
  sample preparation,
    body fluids, 89
    circulating cells, 91
    cultured cells, 91
    general considerations, 88–89
    plant tissues, 91
    solid tissues, 90
  sample solubilization, 91–93
  second dimension (SDS–PAGE),
    method, 102
    molecular mass standards, 102
    transfer between dimensions, 101–102

Urea,
  protein denaturant, 41–42
  isoelectric focusing, 67
  protein modification by cyanate ions, 42, 89
  removal of cyanate ions, 42
  transverse gradient gels, 42–43

Western blotting,
  amino acid analysis, 155
  apparatus,
    semi-dry, 143–144
    vertical tank, 142
  background, 139–141
  blocking, 147–148
  blotting membranes, 144–145
  electroblotting method, 142–144
  general protein detection methods, 146–147
  internal protein sequence analysis, 161
  molecular mass standards, 147
  multiple transfers, 144
  N-terminal protein sequence analysis, 156–159
  principles of method, 140–141
  quantitation, 152
  specific detection,
    avidin–biotin methods, 149

Western blotting—*cont.*
   chemiluminescence, 150–153
   enzyme-conjugated antibodies,
     148–150
   fluorescent antibodies, 148–149
   immunogold technique, 149
   principles of method, 148–149

Western blotting—*cont.*
   protein A, 148
   protein G, 148
   transfer buffers, 145–146
   transfer methods, 141–144

Zone electrophoresis, 9